My first auto[graph]

Granville Sewell

Analysis of a Finite Element Method

Granville Sewell

Analysis of a
Finite Element Method
PDE/PROTRAN

With 32 Illustrations

Springer-Verlag
New York Berlin Heidelberg Tokyo

Granville Sewell
University of Texas at El Paso
Mathematics Department
El Paso, TX 79968
U.S.A.

AMS Classification: 65LXX, 65MXX, 65NXX

Library of Congress Cataloging-in-Publication Data
Sewell, Granville.
  Analysis of a finite element method—PDE/
PROTRAN.
  1. Finite element method—Data processing.
2. Differential, equations, Partial—Data processing.
I. Title.
TA347.F5S49   1985   515.3′53   85-22198

Printed and bound by R.R. Donnelley and Sons, Harrisonburg, Virginia.
Printed in the United States of America.

9 8 7 6 5 4 3 2 1

ISBN 0-387-96226-3 Springer-Verlag New York Berlin Heidelberg Tokyo
ISBN 3-540-96226-3 Springer-Verlag Berlin Heidelberg New York Tokyo

To my family: Melissa, Christopher, and Kevin

Preface

This text can be used for two quite different purposes. It can be used as a reference book for the PDE/PROTRAN user who wishes to know more about the methods employed by PDE/PROTRAN Edition 1 (or its predecessor, TWODEPEP) in solving two-dimensional partial differential equations.

However, because PDE/PROTRAN solves such a wide class of problems, an outline of the algorithms contained in PDE/PROTRAN is also quite suitable as a text for an introductory graduate level finite element course. Algorithms which solve elliptic, parabolic, hyperbolic, and eigenvalue partial differential equation problems are presented, as are techniques appropriate for treatment of singularities, curved boundaries, nonsymmetric and nonlinear problems, and systems of PDEs. Direct and iterative linear equation solvers are studied.

Although the text emphasizes those algorithms which are actually implemented in PDE/PROTRAN, and does not discuss in detail one- and three-dimensional problems, or collocation and least squares finite element methods, for example, many of the most commonly used techniques are studied in detail. Algorithms applicable to general problems are naturally emphasized, and not special purpose algorithms which may be more efficient for specialized problems, such as Laplace's equation. It can be argued, however, that the student will better understand the finite element method after seeing the details of one successful implementation than after seeing a broad overview of the many types of elements, linear equation solvers, and other options in existence.

This text assumes some knowledge of multivariate calculus, linear algebra, and general numerical analysis, such as might be obtained in an introductory undergraduate level course in each of these subjects.

Although PDE/PROTRAN solves systems of PDEs, and the notation reflects this throughout the text, this need not confuse the reader who finds it easier to think in terms of a single PDE. The formulas are always written in such a way that this can be done simply by ignoring the vector (**boldface**) and matrix notation.

There are examples and exercises involving PDE/PROTRAN usage, and the availability of PDE/PROTRAN would make it easy for the student to test the algorithms studied, which otherwise could be tested only after a great amount of programming effort. However, its availability is not essential to the course.

PDE/PROTRAN is available from:

> IMSL, Inc.
> 2500 ParkWest Tower 1
> 2500 CityWest Blvd.
> Houston, TX 77042.

A self-documenting interactive program, INTERPDE, which guides the user through the construction of a PDE/PROTRAN input program is available from:

> Granville Sewell
> Box 12671
> El Paso, TX 79913.

A version (INTERTWO) is also available which is compatible with TWODEPEP.

I would like to thank Lai Tom of IMSL for secretarial support in the preparation of the manuscript.

I would also like to thank Bob Lynch of Purdue, not only for his review of this text, but also for his gift of a book, *An Analysis of the Finite Element Method,* by Gilbert Strang and George Fix, which was the source of so many ideas during the development of TWODEPEP (and PDE/PROTRAN) that the title to my book was chosen to credit this source.

Granville Sewell

Contents

CHAPTER 1. PARTIAL DIFFERENTIAL EQUATION APPLICATIONS

A study of partial differential equation applications gives essential insights into the mathematical properties of the PDEs themselves, and thus the applications discussed in this chapter should not be viewed as supplemental material, but as an integral part of the text.

The applications treated here will be classified according to how they are derived from physical principles. There are a number of ways in which partial differential equations may be derived, but many of the most important applications can be derived using one of three principles: energy minimization, mass balance, and force balance.

1.1 Energy Minimization

If it is possible to express the energy of a system in the form:

$$E(\mathbf{u}) = \iint_R E_1(x,y,\mathbf{u},\mathbf{u}_x,\mathbf{u}_y) \, dxdy + \int_{\partial R_2} E_2(s,\mathbf{u}) \, ds$$

where R is a two dimensional region and ∂R_2 is a portion of its boundary (because PDE/PROTRAN solves two dimensional problems, only two dimensional models will be considered here), and $\mathbf{u}$ is a vector of state variables which must take known values $\mathbf{u}=\mathbf{FB}(x,y)$ on the remaining portion ∂R_1 of the boundary, then a PDE system may be derived by assuming that the steady state condition minimizes the energy of the system.

If $\mathbf{u}$ is the function which minimizes this energy integral, then $E(\mathbf{u}+\epsilon\boldsymbol{\phi})$, as a function of ϵ, must have a minimum at $\epsilon=0$ if $\boldsymbol{\phi}$ is any vector function satisfying $\boldsymbol{\phi}=\mathbf{0}$ on ∂R_1. Thus:

$$0 = dE/d\epsilon(\epsilon=0) =$$

$$\iint_R \{(\partial E_1/\partial\mathbf{u})^T\boldsymbol{\phi} + (\partial E_1/\partial\mathbf{u}_x)^T\boldsymbol{\phi}_x + (\partial E_1/\partial\mathbf{u}_y)^T\boldsymbol{\phi}_y\} \, dxdy + \int_{\partial R_2} (\partial E_2/\partial\mathbf{u})^T\boldsymbol{\phi} \, ds$$

where $\partial E_1/\partial\mathbf{u}$ is used to represent the gradient of E_1 with respect to $\mathbf{u}$ and similarly for the other terms.

To simplify the notation, and to conform more closely to the PDE/PROTRAN notation, the following will be substituted:

(1.1.1) $\mathbf{F} = -\partial E_1/\partial\mathbf{u}$ $\mathbf{A} = \partial E_1/\partial\mathbf{u}_x$ $\mathbf{B} = \partial E_1/\partial\mathbf{u}_y$ $\mathbf{GB} = -\partial E_2/\partial\mathbf{u}$

Using this notation:

$$0 = \iint_R \{\mathbf{F}^T\boldsymbol{\phi} - \mathbf{A}^T\boldsymbol{\phi}_x - \mathbf{B}^T\boldsymbol{\phi}_y\} \, dxdy + \int_{\partial R_2} \mathbf{GB}^T\boldsymbol{\phi} \, ds$$

or, after integration by parts (Green's theorem):

$$0 = \iint_R (\mathbf{A}_x+\mathbf{B}_y+\mathbf{F})^T\boldsymbol{\phi} \, dxdy + \int_{\partial R_2} (\mathbf{GB}-\mathbf{A}n_x-\mathbf{B}n_y)^T\boldsymbol{\phi} \, ds$$

where (n_x, n_y) is the unit outward normal to the boundary.

Since ϕ is an arbitrary function on ∂R_2 and in R, this implies that:

$$0 = A_x(x,y,u,u_x,u_y) + B_y(x,y,u,u_x,u_y) + F(x,y,u,u_x,u_y) \qquad \text{in R}$$

and

(1.1.2)
$$An_x + Bn_y = GB(x,y,u) \qquad \text{on } \partial R_2$$

as well as

$$u = FB(x,y) \qquad \text{on } \partial R_1$$

Some examples follow.

Example 1

The energy of an elastic membrane stretched over a frame is given by:

$$E(u) = \iint_R \{T(\,1+u_x^2+u_y^2\,)^{0.5} - fu\} \, dxdy$$

where $u(x,y)$ is the height of the membrane above a given point in the x-y plane, and R is the projection of the membrane on that plane. The first term in the integral is due to the increase in membrane area caused by stretching (T=tension) and the second is due to the work required to move the membrane a distance u against a vertical force, $f(x,y)$. (Note that potential energy increases if u is in a direction opposing the force.)

According to (1.1.2), using definitions (1.1.1), the solution satisfies:

$$(Tu_x/(1+u_x^2+u_y^2)^{0.5})_x + (Tu_y/(1+u_x^2+u_y^2)^{0.5})_y + f = 0$$

u is also required to equal the height of the frame on the part ∂R_1 of the boundary where it is attached to the frame. On the free part, by (1.1.2), $u_x n_x + u_y n_y = 0$, or $\partial u / \partial n = 0$.

If u_x and u_y are small, this nonlinear equation may be approximated by the linear equation:

$$(Tu_x)_x + (Tu_y)_y + f = 0$$

Example 2

The energy of a two dimensional elastic body subject to in-plane forces is of the form:

$$E(u_1,u_2) = \iint_R E_1(\varepsilon_{11},\varepsilon_{12},\varepsilon_{22},u_1,u_2) \, dxdy + \int_{\partial R_2} E_2(u_1,u_2) \, ds$$

where $\varepsilon_{11} = (u_1)_x$, $\varepsilon_{22} = (u_2)_y$ and $\varepsilon_{12} = (u_1)_y + (u_2)_x$ are the "strains" and (u_1,u_2) is the displacement vector.

Now $\partial E_1 / \partial \varepsilon_{11}$ is by definition σ_{11}, the stress which opposes the strain ε_{11}. That is, σ_{11} is the rate at which potential energy increases with increasing strain ε_{11}. Similarly:

$$\partial E_1/\partial \varepsilon_{12} = \sigma_{12} \qquad \partial E_1/\partial \varepsilon_{22} = \sigma_{22}$$

Further, $\partial E_1/\partial u_1$ is by definition $-f_1$, the force which opposes the displacement u_1 (f_1 is considered positive when in the direction of positive u_1, thus, the opposing force is $-f_1$). Also,

$$\partial E_1/\partial u_2 = -f_2 \ , \ \partial E_2/\partial u_1 = -gb_1 \ , \ \partial E_2/\partial u_2 = -gb_2$$

where (f_1,f_2) represents the body force vector and (gb_1,gb_2) the boundary force vector. Thus (1.1.2) and (1.1.1) give:

$$(\sigma_{11})_x + (\sigma_{12})_y + f_1 = 0 \qquad \text{in R}$$

(1.1.3)

$$(\sigma_{12})_x + (\sigma_{22})_y + f_2 = 0$$

and

$$\sigma_{11}n_x + \sigma_{12}n_y = gb_1 \qquad \text{on } \partial R_2$$

$$\sigma_{12}n_x + \sigma_{22}n_y = gb_2$$

while on the other part of the boundary, ∂R_1, the displacements are assumed to be known:

$$u_1 = fb_1 \qquad \text{on } \partial R_1$$

$$u_2 = fb_2$$

The experimental relationships between the stresses and strains in a three dimensional, linear, anisotropic elastic body can be reduced to two dimensions by assuming either that all out-of-plane stresses are zero ("plane stress", usually assumed to hold in thin plates) or that all out-of-plane strains are zero ("plane strain", usually assumed in thick plates). In the plane stress case, these "constitutive" relations are:

$$\sigma_{11} = E(\varepsilon_{11}+\nu\varepsilon_{22}) \ / \ (1-\nu^2)$$

(1.1.4)

$$\sigma_{12} = 0.5E\varepsilon_{12} \ / \ (1+\nu)$$

$$\sigma_{22} = E(\varepsilon_{22}+\nu\varepsilon_{11}) \ / \ (1-\nu^2)$$

and in the plane strain case:

$$\sigma_{11} = E((1-\nu)\varepsilon_{11}+\nu\varepsilon_{22}) \ / \ (1+\nu) \ / \ (1-2\nu)$$

(1.1.5)

$$\sigma_{12} = 0.5E\varepsilon_{12} \ / \ (1+\nu)$$

$$\sigma_{22} = E((1-\nu)\varepsilon_{22}+\nu\varepsilon_{11}) \ / \ (1+\nu) \ / \ (1-2\nu)$$

where E is the elastic modulus and ν is the Poisson ratio.

Example 3

The bending energy in a rigid plate is modeled as:

$$E(u) = \iint_R (0.5D(\nabla^2 u)^2 - fu)\, dxdy$$

where D is the modulus of rigidity, f is the vertical load, and u is the vertical displacement. In the first term, $\nabla^2 u$ is a measure of the bending, and the second term is the work required to move against the opposing external force.

Although the appearance of second derivatives prevent the application of (1.1.2), the same procedure may be applied to force $E(u+\varepsilon\phi)$ to have a minimum at $\varepsilon=0$. This leads to:

$$0 = \iint_R (\nabla^2(D\nabla^2 u)-f)\phi\, dxdy + \int_{\partial R_2} \{D\nabla^2 u(\partial\phi/\partial n)-\phi\, \partial(D\nabla^2 u)/\partial n\}\, ds$$

If u and $\partial u/\partial n$ are given on the boundary (clamped boundary condition), ϕ and $\partial\phi/\partial n$ are zero there also. If u is given and $\nabla^2 u$ is zero (simply supported boundary) then ϕ is zero on the boundary.

In either case, the boundary integrals disappear, while in R the arbitrariness of ϕ forces:

$$\nabla^2(D\nabla^2 u) = f \qquad \text{in R}$$

It should be pointed out that the introduction of the auxiliary variable $v=D\nabla^2 u$ reduces this fourth order equation to the two second order equations:

$$\nabla^2 u = v/D$$

$$\nabla^2 v = f$$

Elastic shell problems {see 1}, and many other elasticity problems can be formulated as energy minimization problems.

It will be seen later (Section 2.1) that those applications which can be formulated as energy minimization problems are ideally suited for finite element solution and error analysis.

1.2 **Mass Balance**

Let $u(x,y,t)$ represent the concentration (density) of a substance in a
(solid, liquid or gas) medium, at position x,y and time t. Then a PDE
satisfied by u can be derived by balancing the mass entering and leaving an
arbitrary small subregion E of R. If $\mathbf{J}$ represents the flux vector, that is,
a vector such that $\mathbf{J}^T\mathbf{v}$ gives the mass of u crossing a plane perpendicular
to the unit vector $\mathbf{v}$ per unit area per unit time, then the net change in u
per unit time in the subregion E due to mass crossing the boundary is:

$$\int_{\partial E} -\mathbf{J}^T\mathbf{n}\ ds$$

Here $\mathbf{n}$ is the unit outward normal to the boundary of E, so that $-\mathbf{J}^T\mathbf{n}$ is the
inward flux. If $f(x,y,t,u)$ represents the rate at which u is being created
(negative if destroyed) then the net change in u per unit time due to
sources and sinks is $\iint_E f\ dxdy$. Thus:

$$\frac{d}{dt}\iint_E u\ dxdy = \int_{\partial E} -\mathbf{J}^T\mathbf{n}\ ds + \iint_E f\ dxdy$$

or, applying the divergence theorem:

$$\iint_E u_t\ dxdy = \iint_E (-\mathbf{V}^T\mathbf{J} + f)\ dxdy$$

This can be true in an arbitrary subregion E only if:

$$u_t = -\mathbf{V}^T\mathbf{J} + f \qquad \text{in R}$$

If there is anisotropic (direction independent) diffusion, the flux is in
the opposite direction from the gradient of u, that is, it is in the direction
of most rapid decrease in concentration. Thus $\mathbf{J} = -D\mathbf{V}u$. If there is
also convection present, with a convection ("wind" or "current") velocity $\mathbf{v}$,
then the flux has another term, $\mathbf{J} = u\mathbf{v}$. Thus in the presence of both
anisotropic diffusion and convection, the PDE becomes:

$$u_t = \mathbf{V}^T(D\mathbf{V}u - u\mathbf{v}) + f$$

or

$$u_t = (Du_x - uv_1)_x + (Du_y - uv_2)_y + f$$

This PDE is said to be "parabolic", and the initial concentration $u(x,y,0)$
must be specified at t=0 throughout the region R. When steady state conditions are reached, $u_t=0$, and the problem becomes "elliptic":

$$0 = (Du_x - uv_1)_x + (Du_y - uv_2)_y + f$$

If, in the time dependent problem, the diffusion is negligibly slow compared
to the convection, the problem becomes "hyperbolic":

(1.2.1) $$u_t = (-uv_1)_x + (-uv_2)_y + f$$

In any of these cases, the usual boundary conditions are to specify either the value of the concentration u, or the boundary flux $\mathbf{J}^T\mathbf{n}$. This latter condition, in the presence of diffusion and convection, becomes:

$$(Du_x - uv_1)n_x + (Du_y - uv_2)n_y = gb$$

or

$$D\ \partial u/\partial n - u\mathbf{v}^T\mathbf{n} = gb$$

Heat conduction may be considered a special case of diffusion, where the diffusing component is heat energy, given by $c\rho T$, where c = specific heat, ρ = density, T = temperature. The heat equation becomes:

$$c\rho T_t = \mathbf{\nabla}^T(Dc\rho\mathbf{\nabla}T - c\rho T\mathbf{v}) + f$$

The quantity $Dc\rho$ is called the coefficient of thermal conductivity, and is usually denoted by K. So:

$$c\rho T_t = \mathbf{\nabla}^T(K\mathbf{\nabla}T - c\rho T\mathbf{v}) + f$$

If $\mathbf{v}$ represents a divergence free velocity field, $\mathbf{\nabla}^T\mathbf{v} = 0$, then this can alternatively be written:

(1.2.2) $$c\rho T_t = \mathbf{\nabla}^T(K\mathbf{\nabla}T) - c\rho\mathbf{v}^T\mathbf{\nabla}T + f$$

1.3 **Force Balance**

In Section 1.1, Example 1, it was seen that, assuming small displacements, the equation governing the vertical displacement of a membrane at rest is:

$$(Tu_x)_x + (Tu_y)_y + f = 0$$

where f represents the applied vertical force.

If the membrane is not at rest, it may be considered that there is an additional inertial force, equal to the mass (per unit area) times the acceleration, in a direction opposing the acceleration. In addition, there may also be a frictional (or damping) force present which is proportional to, and in an opposing direction from, the velocity. Thus the equation becomes (ρ=density):

$$(Tu_x)_x + (Tu_y)_y + f - \rho u_{tt} - a u_t = 0$$

or

$$(1.3.1) \qquad \rho u_{tt} + a u_t = (Tu_x)_x + (Tu_y)_y + f$$

This equation is "hyperbolic", and since it involves a second time derivative, both the initial position of the membrane $u(x,y,0)$ and its initial velocity $u_t(x,y,0)$ must be specified. Boundary conditions may be the same as for the steady state case.

For future reference (Section 1.5), notice that the change of variable $v = u$ can be used to reduce this equation to two equations which are first order in time:

$$u_t = v$$

$$\rho v_t = -av + (Tu_x)_x + (Tu_y)_y + f$$

with initial conditions on u and $v(=u_t)$.

In a similar manner, the applied forces in the plane stress or plane strain elasticity problem may be considered to include inertial terms opposing the acceleration:

$$\rho(u_1)_{tt} = (\sigma_{11})_x + (\sigma_{12})_y + f_1$$

$$(1.3.2)$$

$$\rho(u_2)_{tt} = (\sigma_{12})_x + (\sigma_{22})_y + f_2$$

Again, this hyperbolic system requires initial conditions on (u_1,u_2) and $((u_1)_t,(u_2)_t)$ and boundary conditions as in the steady state case. Also notice that this system can be written as a system of four equations which are of first order in time, by the introduction of variables $v_1=(u_1)_t$, $v_2=(u_2)_t$.

Thus the force vector accelerating a small unit volume of an elastic medium is $((\sigma_{11})_x+(\sigma_{12})_y+f_1 \ , \ (\sigma_{12})_x+(\sigma_{22})_y+f_2)$, which includes external body forces and stress induced internal forces. In a similar manner it is possible to show that the force acting on a small, moving, unit volume of a fluid is given by this same vector, and that for isotropic, incompressible fluids:

$$\sigma_{11} = -p + 2\mu u_x \ , \quad \sigma_{22} = -p + 2\mu v_y \ , \quad \sigma_{12} = \mu(u_y+v_x)$$

where p is the hydrostatic pressure, μ is the coefficient of viscosity, and (u,v) is the velocity vector. The equations of motion for a fluid are obtained by setting this force equal to the mass (per unit volume) times the acceleration of that **moving** unit volume:

$$\rho \ d/dt(u(x,y,t),v(x,y,t)) = \rho(u_x x_t+u_y y_t+u_t \ , \ v_x x_t+v_y y_t+v_t)$$

Notice that the time derivative of velocity is calculated taking into account the fact that the unit volume is moving, so that x and y are functions of t. Since also $x_t=u$ and $y_t=v$, the equations for the fluid may be written:

$$\rho(u_t + uu_x + vu_y) = (\sigma_{11})_x + (\sigma_{12})_y + f_1$$
$$(1.3.3)$$
$$\rho(v_t + uv_x + vv_y) = (\sigma_{12})_x + (\sigma_{22})_y + f_2$$

These are called the Navier-Stokes equations. For the steady state problem $u_t=v_t=0$.

The boundary conditions are generally either:

$$\sigma_{11}n_x + \sigma_{12}n_y = gb_1$$

$$\sigma_{12}n_x + \sigma_{22}n_y = gb_2$$

which are obtained by balancing the internal (stress induced) forces against the external boundary forces (called tractions), (gb_1,gb_2); or else the velocity on the boundary is given: $u=fb_1, v=fb_2$.

The density (ρ) satisfies the transport equation (1.2.1) with zero source term, which can be rewritten in the form:

$$\rho_t + u\rho_x + v\rho_y = -\rho(u_x+v_y)$$

The left hand side (see above) is the time derivative of the density of a moving volume of fluid, and is therefore zero for an incompressible fluid. This leads to the "divergence" equation $u_x+v_y=0$.

Thus the equations for incompressible fluid flow consist of the Navier-Stokes equations plus the divergence equation. The three unknowns are u,v and the pressure p.

These three equations are difficult to deal with as they stand, particularly because of the lack of boundary conditions on p. One approach to make them more tractable is to assume a "stream function" ϕ such that $u=\phi_y$, $v= -\phi_x$, so that the divergence equation is automatically satisfied since $\phi_{yx}-\phi_{xy} = 0$.

In fact the divergence equation assures that such a stream function exists. Then the bothersome pressure is eliminated from the Navier-Stokes equations by differentiating the first equation with respect to y and subtracting the second equation differentiated with respect to x. (It is now assumed that ρ and μ are constants.) With the aid of the auxiliary "vorticity" variable ω = $-\phi_{xx}-\phi_{yy}$ the Navier-Stokes equations reduce to:

$$0 = \phi_{xx} + \phi_{yy} + \omega$$

(1.3.4)

$$\rho\omega_t = \mu(\omega_{xx} + \omega_{yy}) + \rho(\phi_x\omega_y - \phi_y\omega_x) + (f_2)_x - (f_1)_y$$

An alternative approach, called the "penalty method", assumes that the fluid is not completely incompressible, so that the divergence equation no longer holds exactly, but that it requires a very large increase in hydrostatic pressure, p, to obtain a small increase in density, so that the left hand side of the density equation above is no longer exactly zero, but proportional to p. That is, $p = -\alpha(u_x+v_y)$, where α is very large. This eliminates p and the divergence equation at the same time and it could even be argued that the resulting equations are closer to reality than the incompressibility assumption, if α is chosen realistically. In practice, however, it is not important to choose α realistically; any large number will do, provided it is not so large as to lead to an ill-conditioned matrix.

For either approach, initial conditions (on ϕ and ω or u and v) must be given, for the time dependent problem.

1.4 Resonance

If the vertical load and the boundary displacements in the time dependent membrane problem (1.3.1) are time-periodic with frequency ω, the problem can be written:

$$\rho u_{tt} + a u_t = (Tu_x)_x + (Tu_y)_y + f(x,y)\sin(\omega t) \qquad \text{in } R$$

$$u = fb(x,y)\sin(\omega t) \qquad \text{on } \partial R$$

(1.4.1)

$$u = u0(x,y) \qquad \text{at } t=0$$
$$u_t = u1(x,y)$$

If the damping coefficient (a) is very small, but positive (it will never quite be zero in reality), the effects of the initial conditions will die out slowly, but once they do disappear the solution will be periodic with the same frequency and can be written $u=A(x,y)\sin(\omega t)$ (Problem 1.5). The amplitude distribution is determined by solving the steady state equation which results when this form for u is inserted into the above equation, with a considered negligible despite its role in damping out the transient solutions:

$$-\omega^2\rho A = (TA_x)_x + (TA_y)_y + f(x,y) \qquad \text{in } R$$

(1.4.2)

$$A = fb(x,y) \qquad \text{on } \partial R$$

Normally this problem will have a unique solution, but if ω happens to be an eigenvalue of the homogeneous problem, that is, if:

$$0 = (TA_x)_x + (TA_y)_y + \omega^2\rho A \qquad \text{in } R$$

(1.4.3)

$$A = 0 \qquad \text{on } \partial R$$

has a nonzero solution, then any multiple of this eigenfunction can be added to a solution of (1.4.2) and another solution results. This can interpreted to mean that as ω approaches an eigenvalue of (1.4.3), the problem (1.4.2) becomes nearly singular, and has a solution which becomes very large, so that the amplitude of the vibration increases without bounds. This eigenvalue problem will have an infinite sequence of eigenvalues, called "resonant frequencies", but generally only the smallest ones are of interest.

The time dependent elasticity problem (1.3.2) also has resonant frequencies which are the eigenvalues of:

$$(\sigma_{11})_x + (\sigma_{12})_y + \omega^2\rho A = 0$$
$$\qquad \text{in } R$$
$$(\sigma_{12})_x + (\sigma_{22})_y + \omega^2\rho B = 0$$

with
$$\sigma_{11}n_x + \sigma_{12}n_y = 0$$
$$\qquad \text{on } \partial R_2$$
$$\sigma_{12}n_x + \sigma_{22}n_y = 0$$

and

$$A = 0$$
$$B = 0$$

on ∂R_1

where (A,B) is the displacement amplitude vector of the standing wave, and $\sigma_{11}, \sigma_{12}, \sigma_{22}$ above are homogeneous, linear functions of A and B and their derivatives.

Another important eigenvalue application is the Schrodinger equation of quantum mechanics:

$$(u_x)_x + (u_y)_y - KP(x,y)u + \lambda Ku = 0$$

where K is a known constant, $P(x,y)$ is the potential energy distribution, λ is the eigenvalue which represents the energy level, and $u(x,y)$ is the corresponding eigenfunction, which gives the probability amplitude.

1.5 PDE/PROTRAN

It should be noted that all of the steady state applications discussed
in the preceding sections can be written in the form:

$$0 = A_x(x,y,u,u_x,u_y) + B_y(x,y,u,u_x,u_y) + F(x,y,u,u_x,u_y) \quad \text{in } R$$

$$(1.5.1) \qquad\qquad u = FB(x,y) \qquad\qquad \text{on } \partial R_1$$

$$An_x + Bn_y = GB(x,y,u) \qquad\qquad \text{on } \partial R_2$$

The time dependent applications can all be written in the form:

$$C(x,y,t,u)u_t = A_x(x,y,t,u,u_x,u_y) + B_y(x,y,t,u,u_x,u_y) + F(x,y,t,u,u_x,u_y) \quad \text{in } R$$

$$u = FB(x,y,t) \qquad\qquad \text{on } \partial R_1$$

$$(1.5.2) \qquad An_x + Bn_y = GB(x,y,t,u) \qquad\qquad \text{on } \partial R_2$$

$$u = UO(x,y) \qquad\qquad \text{at } t=t_0$$

----

where C is a diagonal matrix. Finally, all the eigenvalue applications
can be written in the form:

$$0 = A_x(x,y,u,u_x,u_y) + B_y(x,y,u,u_x,u_y) + F(x,y,u,u_x,u_y) + \lambda P(x,y)u \quad \text{in } R$$

$$(1.5.3) \qquad\qquad u = 0 \qquad\qquad \text{on } \partial R_1$$

$$An_x + Bn_y = GB(x,y,u) \qquad \text{on } \partial R_2$$

----

where **A,B,F** and **GB** are linear homogeneous functions, and P is a matrix.

PDE/PROTRAN can formally solve each of the three problems 1.5.1-3. Chapters
2-3 outline the techniques used by PDE/PROTRAN to solve the steady state
system 1.5.1, Chapter 4 outlines the techniques to solve the time dependent
problem 1.5.2, and Chapter 6 describes the solution of problem 1.5.3.

PDE/PROTRAN employs a sophisticated preprocessor to read the user's input program and translate it into FORTRAN. Thus the input can be written in a very high level language, consisting of keywords and keyword phrases which describe the problem and solution options in a simple, readable form. In addition, the preprocessor does a large amount of input error checking (see Appendix 2).

To illustrate the PDE/PROTRAN input format, consider the following two examples.

(I)   Steady state elastic crack problem

      The elasticity equations (1.1.3) are solved with plane stress constitu-
      tive relations (1.1.4), in a cracked compact tension specimen (Figure
      1.5.1). The two edges adjacent to the crack are pulled apart, so that
      the boundary conditions consist of applying separating boundary forces
      along these edges and zero boundary forces on the rest of the specimen.
      The PDE/PROTRAN input to solve this problem is shown in Table 1.5.1.
      The output stress distribution is shown in Figure 1.5.1.

(II) Heat conduction-convection problem

      The heat conduction-convection equations (1.2.2) are solved in a
      rectangle, with convection velocity **v** as shown in Figure 1.5.2. The
      boundary conditions specify that the temperature at the top is held at
      0, while the temperature at the bottom is 13. The sides are insulated,
      so that there is zero flux across them. The initial condition specifies
      that temperature is a linear function of y, u = 13(1-y), as would be
      the steady state solution in the absence of convection.

      The PDE/PROTRAN input is shown in Table 1.5.2 and the temperature
      distribution at t=0.1 is shown in Figure 1.5.3.

Detailed descriptions of the input parameters may be found in the PDE/PROTRAN User's Manual (or see Appendix 3), but the language is so self-documenting that these examples are largely understandable without the manual.

                     Table 1.5.1

```
C       ELASTIC CRACK PROBLEM
C
$       PDE2D
        UNKNOWNS = (U,V)
        NTRIANGLES = 250
        FRONTAL
        A = (E1*UX+E2*VY   , E3*(UY+VX) )
        B = (E3*(UY+VX)    , E1*VY+E2*UX)
        DEFINE
        ====
C   EM AND VNU ARE ELASTIC MODULUS AND POISSON RATIO
        EM = 10.6E6
        VNU = 0.3
        E1 = EM/(1.-VNU**2)
        E2 = E1*VNU
```

```
      E3 = EM/(1.+VNU)/2.
      ====
C   LOCALLY REFINE TRIANGULATION NEAR THE CRACK TIP
      TRIDENSITY = ((X-1.46)**2+Y**2)**(-1.25)
      SAVEFILE = PLOT
      SYMMETRIC
C   APPLY BOUNDARY FORCE ABOVE AND BELOW CRACK OPENING
      GB = (1, 0, -1./2.08)  (6, 0,  1./2.08)
C   DEFINE INITIAL TRIANGULATION
      VERTICES = (0,-2.08) (4.25,-2.08) (4.25,0) (4.25,2.08)
    *   (0,2.08) (0,0.001) (0,-0.001) (2,-1) (1.46,0) (2,1)
      TRIANGLES = (7,1,8,1) (1,2,8,2) (2,3,8,3) (3,9,8,0) (9,7,8,4)
    *   (9,3,10,0) (3,4,10,3) (4,5,10,5) (5,6,10,6) (6,9,10,7)
$     PLOTSTRESSES
$     END
```


Table 1.5.2


```
C     HEAT CONDUCTION-CONVECTION PROBLEM
C
$     PDE2D
      C = 1.0
      A = 0.6*UX
      B = 0.6*UY
      F = -V1*UX-V2*UY
      U0 = 13.0*(1.0-Y)
      DEFINE
      ====
C   DEFINE CONVECTION VELOCITY COMPONENTS
      V1 = -31.4159*SIN(3.14159*Y)*SIN(3.14159*X/6.)
      V2 = -31.4159*COS(3.14159*Y)*COS(3.14159*X/6.)/6.0
      ====
C   TEMPERATURE AT BOTTOM (AT TOP DEFAULTS TO FB=0)
      FB = (-2,13.0)
      GRIDPOINTS = (13,13)
      SAVEFILE = PLOT
      OUTFREQUENCY = 2
      CRANKNICOLSON
      DT = 0.01
      TF = 0.1
      NOUPDATE
C   USE RECTANGULAR GRID TO DEFINE INITIAL TRIANGULATION
      XGRID = 0,1.5,3.0,4.5,6.0
      YGRID = 0,0.05,0.25,0.45,0.50,0.55,0.75,0.95,1.0
      ARCS = (1,3) (-2,-4)
$     PLOTSURFACE
      SET = 5
      SURFACE = U
      TITLE = 'TEMPERATURE AT T=0.10'
$     PLOTCONTOUR
      SET = 5
$     END
```

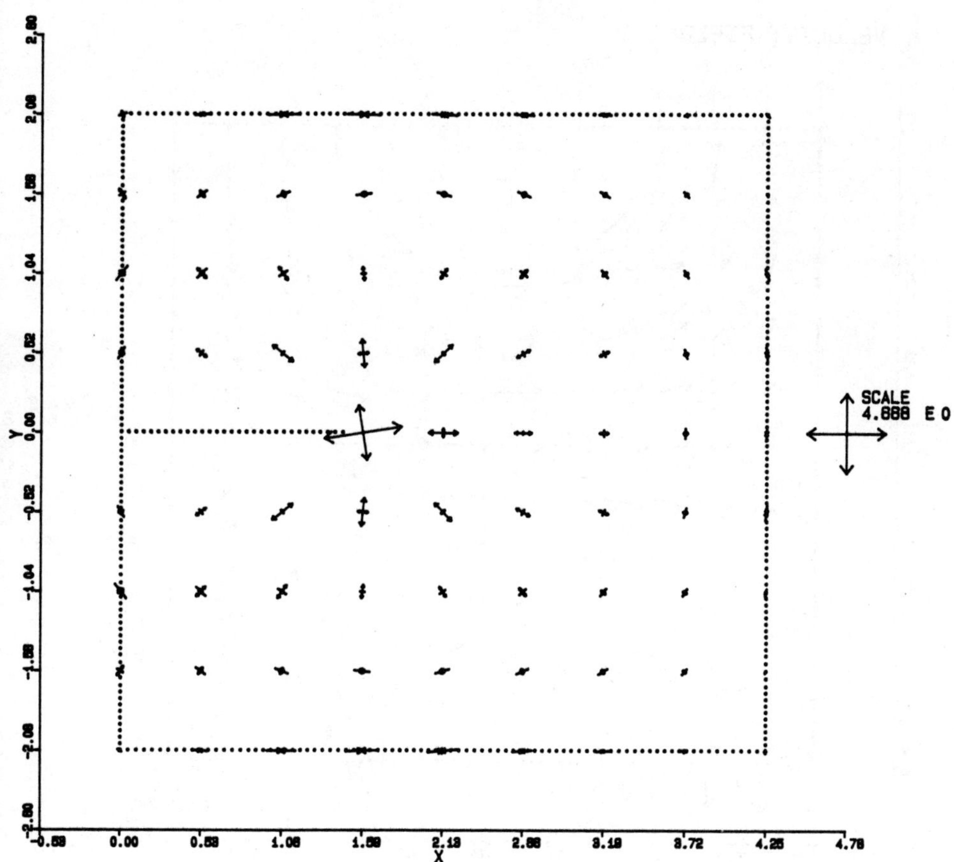

Figure 1.5.1   Stresses in Elastic Crack Problem

15

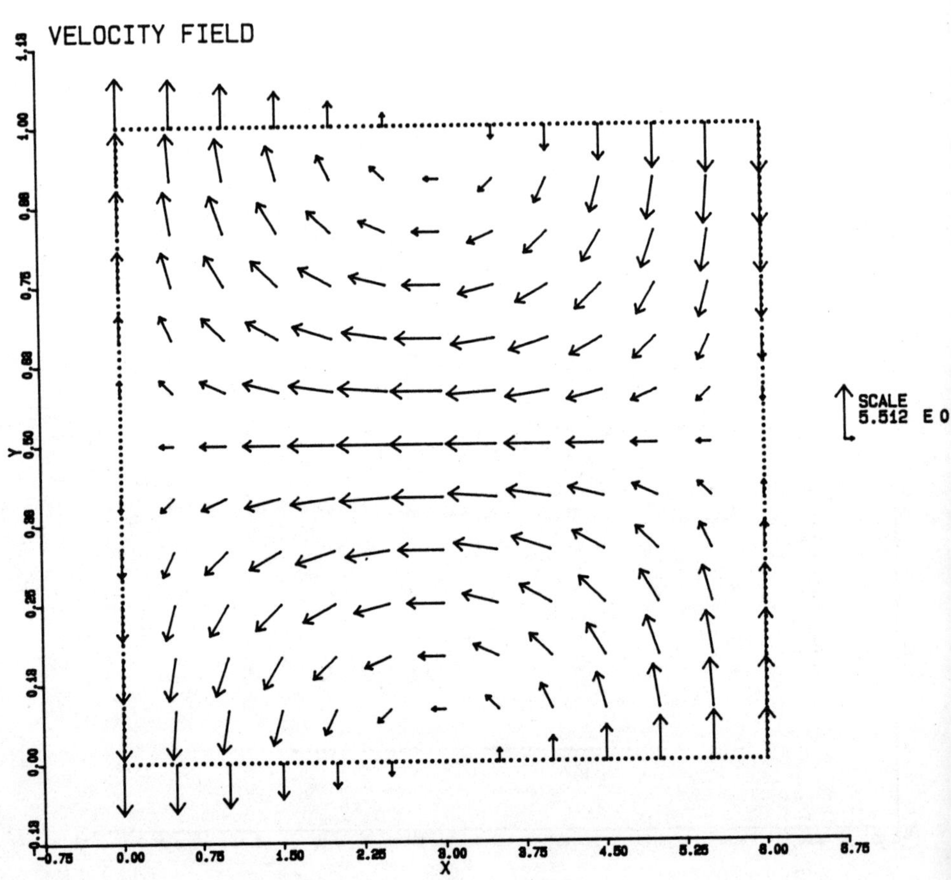

Figure 1.5.2 Velocity Field for Conduction-Convection Problem

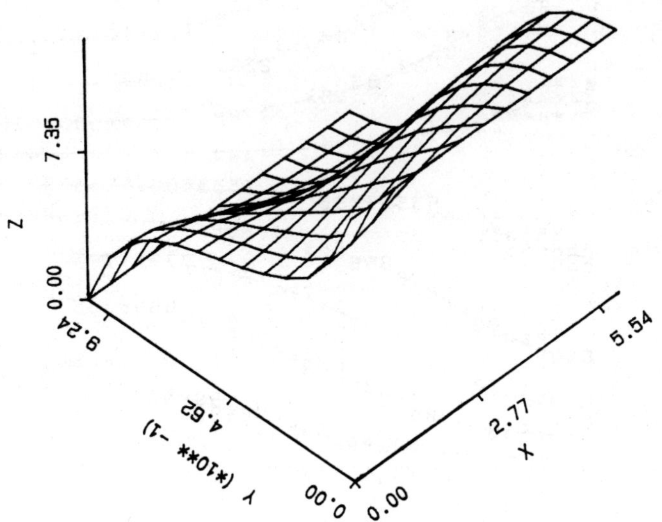

Figure 1.5.3a   Temperature Distribution for Conduction-Convection Problem

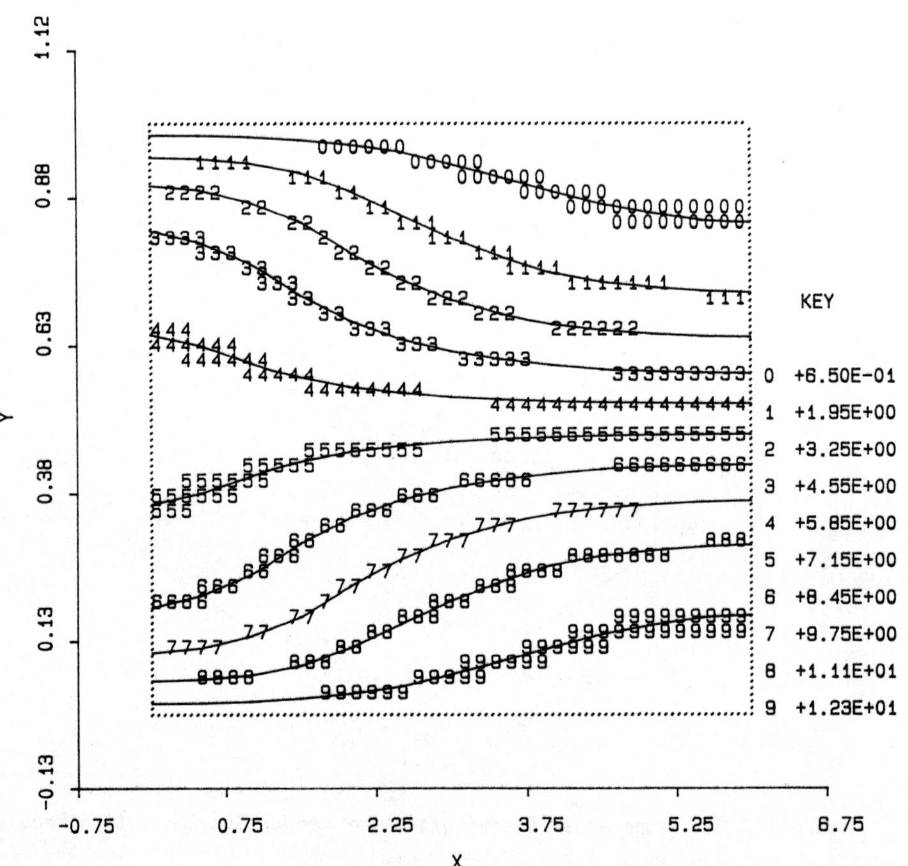

Figure 1.5.3b   Contour Plot of Temperature Distribution

1.6 **Exercises** (* = requires use of PDE/PROTRAN)

1.1 In addition to those mentioned in this chapter, there are still other ways in which PDEs may be derived, as illustrated by the following.

    a. Verify that in any (3D) inverse square force field, the potential energy is proportional to $1/r$. (Hint: force = -gradient of potential energy)

    b. Thus in any inverse square force field (e.g. electrostatic or gravitational) the potential energy at a point (x,y,z) is equal to the sum of the potential energies due to all attracting elements, i.e.:

$$u(x,y,z) = \iint_S \frac{f(\xi,\eta,\zeta)\ d\xi d\eta d\zeta}{((x-\xi)^2+(y-\eta)^2+(z-\zeta)^2)^{0.5}}$$

where $f(\xi,\eta,\zeta)$ is the strength of the source at (ξ,η,ζ), and S is the set of all source points, assumed not to include (x,y,z). Now show, using this formula, that $\nabla^2 u = 0$.

1.2 The constitutive equations for the penalty method formulation of the fluid flow equations are:

$$\sigma_{11} = -p + 2\mu u_x = \alpha(u_x+v_y) + 2\mu u_x$$

$$\sigma_{22} = -p + 2\mu v_y = \alpha(u_x+v_y) + 2\mu v_y$$

$$\sigma_{12} = \mu(u_y+v_x)$$

Show that these are equivalent to the plain strain equations (1.1.5) with $\nu = 0.5/(1+\mu/\alpha)$, $E = 2\mu(1+\nu)$. Thus at low velocities (so that the nonlinear terms are negligible) the steady state Navier-Stokes equations (penalty formulation) are mathematically equivalent to the solid elasticity (plane strain) problem with Poisson ratio almost equal to 0.5.

1.3 In the examples of this chapter, three dimensional problems have been reduced to two dimensional ones by assuming the solution to be independent of the third dimensional variable, z. Two dimensional problems may also result from axial symmetry. For example, show that the equation:

$$u_{xx} + u_{yy} + u_{zz} = 0$$

reduces to

$$(ru_r)_r + (ru_z)_z = 0$$

if u is a function of the cylindrical coordinates r and z, but independent of θ, where:

$$x = r \cos \theta$$
$$y = r \sin \theta$$
$$z = z$$

The PDE/PROTRAN User's Manual shows how to formulate axisymmetric elasticity and fluid flow problems.

1.4 When diffusion is very slow compared to convection, the diffusion-convection equation approaches the hyperbolic problem (1.2.1):

$$u_t = -\mathbf{V}^T(u\mathbf{v}) + f$$

It seems reasonable physically that now the concentration u should be specified only on that part of the boundary where the flux is inward, i.e., where $\mathbf{v}^T\mathbf{n}$ is negative. Assuming an incompressible velocity field ($\mathbf{V}^T\mathbf{v} = 0$) show that with this boundary condition, plus specification of the initial concentration $u(x,y,0)$, the convection problem has (at most) a unique solution. Thus, the solution on the downwind portion of the boundary, where the flux is outward, is automatically specified as part of the solution and must not be given as part of the boundary conditions. (Hint: Let $e=u_1-u_2$, where u_1 and u_2 are solutions, and multiply by e the PDE satisfied by e, and integrate over R.)

1.5 a. Show that if $A(x,y)$ and $B(x,y)$ satisfy the following:

$$-\omega^2\rho A - \omega a B = (TA_x)_x + (TA_y)_y + f(x,y) \qquad \text{in R}$$

$$-\omega^2\rho B + \omega a A = (TB_x)_x + (TB_y)_y$$

$$A = fb(x,y) \qquad\qquad \text{on } \partial R$$
$$B = 0$$

and $e(x,y,t)$ satisfies the following:

$$\rho e_{tt} + a e_t = (Te_x)_x + (Te_y)_y \qquad \text{in R}$$

$$e = 0 \qquad\qquad \text{on } \partial R$$

$$e = u0(x,y)-B(x,y) \qquad\qquad \text{at } t=0$$
$$e_t = u1(x,y) - \omega A(x,y)$$

then $u=A(x,y)\sin(\omega t) + B(x,y)\cos(\omega t) + e(x,y,t)$ solves the initial value problem (1.4.1).

b. Show that the transient solution $e(x,y,t)$ dies out with time, by showing that the energy:

$$\iint\limits_{R} (\rho e_t^2 + T|\mathbf{V}e|^2)\ dxdy$$

is steadily decreasing. (Hint: multiply by e_t the equation satisfied by e, and integrate over R.)

c. If a is negligibly small (although positive to ensure that the transient solution does eventually die out), conclude that B is zero and that A satisfies (1.4.2).

*1.6 Use PDE/PROTRAN to solve the steady state problem:

$$\nabla^2 u + 4 = 0 \qquad \text{in the upper half of the unit circle}$$

with boundary conditions:

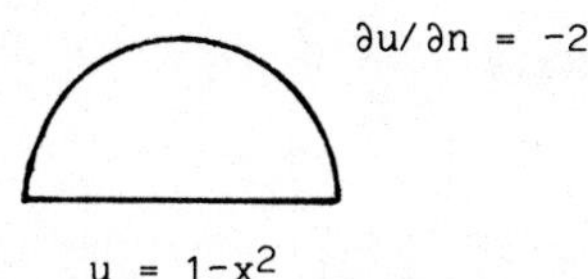

Compare with the exact solution $u=1-x^2-y^2$, and make a surface plot of the solution.

This is a simple problem intended to introduce you to the mechanics of using PDE/PROTRAN. If INTERPDE (see Preface) is available, you can simply answer all of the questions which that interactive program poses, and your PDE/PROTRAN input will be created automatically for you with no further documentation needed. INTERPDE will also work through an example problem with you if you so request. Otherwise, you will need a copy of the PDE/PROTRAN manual to help you construct this input program.

Reference

1.   G. Sewell, "Applications of TWODEPEP," PDE Software: Modules, Interfaces and Systems, edited by B. Engquist and T. Smedsaas, North Holland, 1984, pp 225-240.

CHAPTER 2. ELLIPTIC PROBLEMS - FORMING THE ALGEBRAIC EQUATIONS

2.1 The Galerkin Method

The form of the steady state PDE system solved by PDE/PROTRAN (Section 1.5), is:

$$0 = \mathbf{A}_x(x,y,\mathbf{u},\mathbf{u}_x,\mathbf{u}_y) + \mathbf{B}_y(x,y,\mathbf{u},\mathbf{u}_x,\mathbf{u}_y) + \mathbf{F}(x,y,\mathbf{u},\mathbf{u}_x,\mathbf{u}_y) \quad \text{in } R$$

$$(2.1.1) \qquad\qquad \mathbf{u} = \mathbf{FB}(x,y) \qquad \text{on } \partial R_1$$

$$A n_x + B n_y = \mathbf{GB}(x,y,\mathbf{u}) \qquad \text{on } \partial R_2$$

where R is a general two dimensional region and ∂R_1 and ∂R_2 are disjoint parts of the boundary. The time dependent and eigenvalue problems will be studied in Chapters 4-6.

In this chapter, a finite element discretization of problem (2.1.1) will be developed. It will be observed that much more programming effort is required to obtain the discrete algebraic equations than is required using finite differences, but the effort will be rewarded by high accuracy, even when irregular regions are treated, and even when local mesh refinement is required. In Chapter 3, the algorithms used by PDE/PROTRAN to solve these discrete equations will be studied.

The foundation of the Galerkin finite element method is to choose a set of basis functions $\phi_1 \ldots \phi_N$, for some approximating space, which satisfy $\phi_j = 0$ on ∂R_1 and to approximate the solution of (2.1.1) by:

$$(2.1.2) \qquad \mathbf{u_G}(x,y) = \boldsymbol{\phi}_0(x,y) + \sum_{j=1}^{N} \mathbf{a_j} \, \phi_j(x,y)$$

where $\boldsymbol{\phi}_0(x,y)$ is a function which satisfies $\boldsymbol{\phi}_0 = \mathbf{FB}$ on ∂R_1.

If the PDE and second boundary condition are multiplied by an arbitrary smooth function $\phi(x,y)$ which satisfies $\phi=0$ on ∂R_1, and integrated over R and ∂R_2 respectively:

$$\iint_R (\mathbf{A}_x + \mathbf{B}_y + \mathbf{F})\phi \; dxdy + \int_{\partial R_2} (\mathbf{GB} - A n_x - B n_y)\phi \; ds = \mathbf{0}$$

Or, after integration by parts (Green's theorem):

$$(2.1.3) \quad \iint_R (\mathbf{F}\phi - \mathbf{A}\phi_x - \mathbf{B}\phi_y) \; dxdy + \int_{\partial R_2} \mathbf{GB}\phi \; ds = \mathbf{0}$$

Now (2.1.3) is called the "weak formulation" of (2.1.1). Since ϕ is arbitrary in R and on ∂R_2, requiring (2.1.3) is essentially equivalent to imposing the PDE and second boundary condition.

Since a function $\mathbf{u_G}$ of the form (2.1.2) cannot, in general, be found which satisfies (2.1.3) for any ϕ which is zero on ∂R_1, it will only be required that (2.1.3) be satisfied for $\phi=\phi_1,\ldots\phi_N$:

(2.1.4) $f_k(a_1...a_N) = \iint\limits_R \{F_G\phi_k - A_G(\phi_k)_x - B_G(\phi_k)_y\}\ dxdy + \int\limits_{\partial R_2} GB_G\phi_k\ ds = 0,\ k=1...N$

where **F,A** and **B** are evaluated at $(x,y,\phi_0 + \sum a_j\phi_j,...)$ and **GB** is evaluated at $(x,y,\phi_0 + \sum a_j\phi_j)$, as indicated by the subscript G.

If there are m PDEs in (2.1.1), then (2.1.4) constitutes the system of Nm algebraic (linear or nonlinear) equations for the Nm unknowns **a_j** which must be solved by PDE/PROTRAN.

In the general case, it is expected that requiring (2.1.4) to be satisfied for more and more functions ϕ_k ($N\rightarrow\infty$) will force the PDE and the second (natural) boundary condition to be satisfied more and more closely. (The first--essential--boundary condition is automatically satisfied.)

Suppose, on the other hand, that (2.1.1) can be derived (even if, as for the diffusion equation, it normally is not) as the equation satisfied by the function which minimizes an integral of the form (Section 1.1):

$$E(\mathbf{u}) = \iint\limits_R E_1(\mathbf{u},\mathbf{u}_x,\mathbf{u}_y)\ dxdy + \int\limits_{\partial R_2} E_2(\mathbf{u})\ ds$$

Then the Galerkin method error can be further analyzed. For, let $\mathbf{u_0}$ and $\mathbf{u_1}$ be any two (smooth) functions which satisfy the first boundary condition. Then, calling $\mathbf{e}=\mathbf{u_1}-\mathbf{u_0}$, according to the multivariate Taylor's series with remainder:

(2.1.5) $\quad E(\mathbf{u_1}) = E(\mathbf{u_0}+\mathbf{e}) =$

$\iint\limits_R E_1(\mathbf{u_0}+\mathbf{e},(\mathbf{u_0})_x+\mathbf{e}_x,(\mathbf{u_0})_y+\mathbf{e}_y)\ dxdy + \int\limits_{\partial R_2} E_2(\mathbf{u_0}+\mathbf{e})\ ds =$

$\iint\limits_R \{(E_1)_0 + (\partial E_1/\partial u)_0^T\mathbf{e} + (\partial E_1/\partial u_x)_0^T\mathbf{e}_x + (\partial E_1/\partial u_y)_0^T\mathbf{e}_y$

$\quad + 0.5(\mathbf{e},\mathbf{e}_x,\mathbf{e}_y)^T H_1(\xi)(\mathbf{e},\mathbf{e}_x,\mathbf{e}_y)\}\ dxdy +$

$\int\limits_{\partial R_2} \{(E_2)_0 + (\partial E_2/\partial u)_0^T\mathbf{e} + 0.5\mathbf{e}^T H_2(\eta)\mathbf{e}\}\ ds$

where $H_1(\xi)$ is a 3m by 3m Hessian matrix of second derivatives of E_1, evaluated at some point ξ between $(\mathbf{u_0},(\mathbf{u_0})_x,(\mathbf{u_0})_y)$ and $(\mathbf{u_1},(\mathbf{u_1})_x,(\mathbf{u_1})_y)$, and $H_2(\eta)$ is an m by m Hessian matrix of second derivatives of E_2. The subscript 0 denotes evaluation at $\mathbf{u_0}$.

Using relations (1.1.1) and defining:

$H(\mathbf{e},\mathbf{e}) = \iint\limits_R (\mathbf{e},\mathbf{e}_x,\mathbf{e}_y)^T H_1(\xi)(\mathbf{e},\mathbf{e}_x,\mathbf{e}_y)\ dxdy + \int\limits_{\partial R_2} \mathbf{e}^T H_2(\eta)\mathbf{e}\ ds$

then (2.1.5) can be written:

$(2.1.6) \quad E(\mathbf{u_1}) = E(\mathbf{u_0}) + \iint\limits_{R} (-\mathbf{F_0}^T\mathbf{e} + \mathbf{A_0}^T\mathbf{e_x} + \mathbf{B_0}^T\mathbf{e_y})\ dxdy$

$$- \int\limits_{\partial R_2} \mathbf{GB_0}^T\mathbf{e}\ ds \quad + 0.5\ H(\mathbf{u_1}-\mathbf{u_0},\mathbf{u_1}-\mathbf{u_0})$$

In particular, if $\mathbf{u_0}$ is taken to be the true solution of the PDE system, $\mathbf{u}$, the integral terms in (2.1.6) vanish according to (2.1.3), since $\mathbf{e} = \mathbf{0}$ on ∂R_1, and so:

$(2.1.7) \quad E(\mathbf{u_1}) = E(\mathbf{u}) + 0.5H(\mathbf{u_1}-\mathbf{u},\mathbf{u_1}-\mathbf{u}) \qquad \mathbf{u_1}$ = any function satisfying first boundary condition

On the other hand, if $\mathbf{u_0}$ is taken to be the Galerkin solution (to 2.1.4) and $\mathbf{u_1}$ is taken to be any other function of the form (2.1.2), the integrals in (2.1.6) again vanish, according to (2.1.4), since $\mathbf{e}$ is a linear combination of the basis functions ϕ_k. Thus:

$(2.1.8) \quad E(\mathbf{u_1}) = E(\mathbf{u_G}) + 0.5H(\mathbf{u_1}-\mathbf{u_G},\mathbf{u_1}-\mathbf{u_G}) \qquad \mathbf{u_1}$ = any function of the form (2.1.2)

Now if the PDE and boundary conditions are linear, the Hessian matrices H_1 and H_2 are functions of x and y only. Then $H(\mathbf{e},\mathbf{e})$ is positive for any $\mathbf{e}$ which vanishes on ∂R_1, since by (2.1.7):

$$0 \leq E(\mathbf{u+e}) - E(\mathbf{u}) = 0.5H(\mathbf{e},\mathbf{e})$$

according to the assumption that $\mathbf{u}$ minimizes E. So $(H(\mathbf{e},\mathbf{e}))^{0.5}$ is a norm; call it $||\mathbf{e}||_H$. Then (2.1.8) implies that $\mathbf{u_G}$ minimizes E over the set of all functions in the approximating set, of the form (2.1.2). Therefore if $\mathbf{u_I}$ is any member of this approximating set,

$$E(\mathbf{u_G}) \leq E(\mathbf{u_I})$$

$$E(\mathbf{u_G})-E(\mathbf{u}) \leq E(\mathbf{u_I})-E(\mathbf{u})$$

$$H(\mathbf{u_G}-\mathbf{u},\mathbf{u_G}-\mathbf{u}) \leq H(\mathbf{u_I}-\mathbf{u},\mathbf{u_I}-\mathbf{u})$$

$$||\mathbf{u_G}-\mathbf{u}||_H \leq ||\mathbf{u_I}-\mathbf{u}||_H$$

The final result, then, is that the Galerkin solution, $\mathbf{u_G}$, is, of all functions in the approximating set, the closest to the true solution in the "H-norm."

In the nonlinear case, $H(\mathbf{e},\mathbf{e})$ is still positive provided ξ and η are in a certain neighborhood of $\mathbf{u}$. And if $\mathbf{u_G}$ and $\mathbf{u_I}$ are in this neighborhood, it is still true that:

$$H(\mathbf{u_G}-\mathbf{u},\mathbf{u_G}-\mathbf{u}) \leq H(\mathbf{u_I}-\mathbf{u},\mathbf{u_I}-\mathbf{u})$$

where the ξ and η in the definition of H may be slightly different on the two sides of the the inequality.

The importance of this result in practice is that if it can be shown that a particular member of the approximating set, say an interpolant, can be found which is close to the true solution, then it may be reasonably assumed that the Galerkin solution is of comparable accuracy, even measured in a different norm. In fact, this is generally assumed even in the case that the PDE cannot be derived from an integral minimization principle! See {1, Section 3.4} for a rigorous justification of this assumption in certain linear cases.

Now, since the system of Nm equations (2.1.4) is solved using Newton's method (Section 3.1) it is useful here to display the Jacobian of this system. This Nm by Nm Jacobian may be thought of as an N by N matrix whose "elements" are the m by m Jacobians:

$$(2.1.9) \quad J_{kj} =$$

$$\iint_R (\phi_k,(\phi_k)_x,(\phi_k)_y)^T \begin{bmatrix} F.U & F.UX & F.UY \\ -A.U & -A.UX & -A.UY \\ -B.U & -B.UX & -B.UY \end{bmatrix} \begin{bmatrix} \phi_j \\ (\phi_j)_x \\ (\phi_j)_y \end{bmatrix} dxdy \quad + \int_{\partial R_2} \phi_k\, GB.U\, \phi_j\, ds$$

where F.U, etc., are m by m Jacobian matrices, and are evaluated at $\mathbf{u_G}$.

By comparing J_{jk} and J_{kj} it will be seen that the Nm by Nm Jacobian will be symmetric if and only if the 3m by 3m matrix:

$$(2.1.10) \quad \begin{bmatrix} F.U & F.UX & F.UY \\ -A.U & -A.UX & -A.UY \\ -B.U & -B.UX & -B.UY \end{bmatrix}$$

is symmetric, and the m by m matrix GB.U is symmetric.

In the case of a PDE system derived from an integral minimization principle, it can be seen using the definitions (1.1.1) that these two matrices are simply $-H_1$ and $-H_2$, respectively, so that they are automatically symmetric. Thus the Nm by Nm Jacobian J is symmetric, and it is also negative definite near $\mathbf{u_G}$ since, as shown in Problem 2.1, for any nonzero Nm vector $\mathbf{w}$, if $\mathbf{W} = \sum \mathbf{w_k}\phi_k$, where the $\mathbf{w_k}$ are the (m-vector) components of $\mathbf{w}$:

$$(2.1.11) \quad \mathbf{w}^{*T}J\mathbf{w} = \iint_R (\mathbf{W}^*,\mathbf{W}_x^*,\mathbf{W}_y^*)^T \begin{bmatrix} F.U & F.UX & F.UY \\ -A.U & -A.UX & -A.UY \\ -B.U & -B.UX & -B.UY \end{bmatrix} \begin{bmatrix} \mathbf{W} \\ \mathbf{W}_x \\ \mathbf{W}_y \end{bmatrix} dxdy \quad + \int_{\partial R_2} \mathbf{W}^{*T}GB.U\,\mathbf{W}\, ds$$

and, since $\mathbf{W}$ is real (the complex form is used for future reference), this is just $-H(\mathbf{W},\mathbf{W})$, which is negative. $\mathbf{W}^*$ is the complex conjugate of $\mathbf{W}$.

It is also easy to see that if H_1 and H_2 are everywhere positive definite, then J is negative definite, although this is not a necessary condition.

2.2 Lagrangian Isoparametric Triangular Elements

In the finite element method, the approximating functions $\phi_0, \phi_1 \ldots \phi_N$ in formula (2.1.2) are chosen to be piecewise polynomials, that is, functions which are polynomials within each "element" of a partition of R.

In PDE/PROTRAN the finite "elements" are triangles and the approximating functions are polynomials of degree 2,3 or 4 restricted to each triangle, and are continuous across triangle boundaries. The quadratic (second degree) piecewise polynomials will be studied in detail, and then the cubic and quartic elements will reviewed briefly.

The region R, which will at first be assumed to be polygonal, is divided into triangles in such a way that the edges of adjacent triangles match up, as shown in a typical triangulation in Figure 2.2.1. The vertices and the midpoints of all sides in this triangulation are called "nodes."

The approximating space S_2 is defined as the set of all continuous functions which, in each triangle k, have the form:

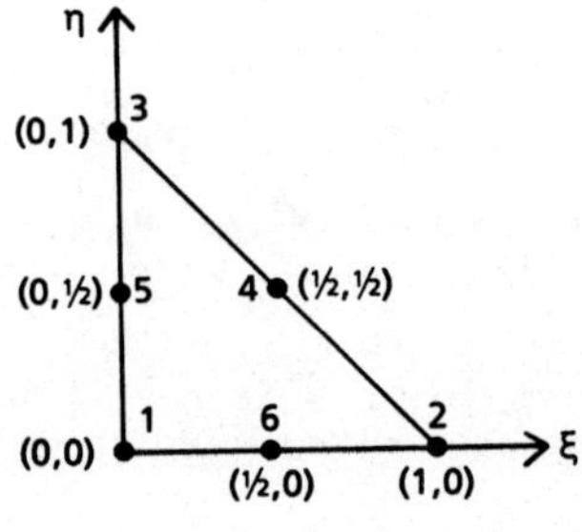

Figure 2.2.1

$$s(x,y) = a_k + b_k x + c_k y + d_k x^2 + e_k xy + f_k y^2$$

In the normalized triangle of Figure 2.2.2, the six unknown coefficients are uniquely determined by specifying the values, $z_1 \ldots z_6$ of $s(\xi,\eta)$ at the six nodes. In fact, $s(\xi,\eta)$ is explicitly given by:

$$s(\xi,\eta) =$$

$$z_1 + (-z_2 + 4z_6 - 3z_1)\xi + (-z_3 + 4z_5 - 3z_1)\eta$$

$$+ (2z_2 - 4z_6 + 2z_1)\xi^2 + (2z_3 - 4z_5 + 2z_1)\eta^2$$

$$+ (4z_1 - 4z_6 + 4z_4 - 4z_5)\xi\eta$$

or

$$s(\xi,\eta) = z_1(1 - 3\xi - 3\eta + 2\xi^2 + 2\eta^2 + 4\xi\eta)$$

$$+ z_2(-\xi + 2\xi^2) + z_3(-\eta + 2\eta^2) + z_4(4\xi\eta)$$

$$+ z_5(4\eta - 4\eta^2 - 4\xi\eta) + z_6(4\xi - 4\xi^2 - 4\xi\eta)$$

or

$$(2.2.1) \qquad s(\xi,\eta) = \sum z_i \psi_i(\xi,\eta)$$

Figure 2.2.2

It is easy to see that the change of variables:

(2.2.2)
$$x = (x_2-x_1)\xi + (x_3-x_1)\eta + x_1$$
$$y = (y_2-y_1)\xi + (y_3-y_1)\eta + y_1$$

maps the six nodes of the normalized triangle (Figure 2.2.2) onto the six nodes of the general triangle of Figure 2.2.3.

Thus the unique function in S_2 which takes the values $z_1...z_6$ at the nodes of the triangle in Figure 2.2.3 is simply the function $s(\xi,\eta)$ after the transformation (2.2.2) has been applied to it, that is,

$$s'(x,y) = s(\xi(x,y),\eta(x,y))$$

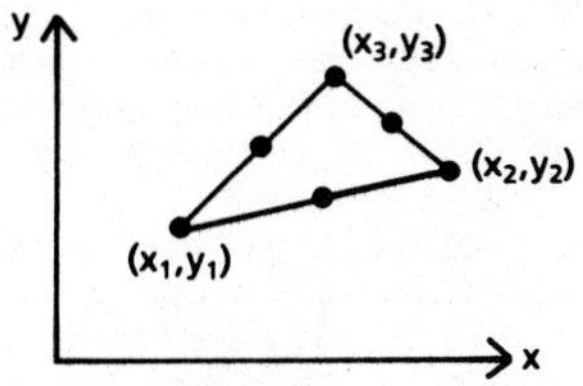

Figure 2.2.3

As for the continuity across triangle boundaries, both neighboring polynomials reduce to quadratics in one variable along any triangle edge, and since they agree at the three nodal points, these two quadratics must be identical.

Now suppose that part of the boundary of R is curved. Then it will require a more complicated transformation to map the normalized triangle onto a triangle adjacent to a curved boundary.

For, consider a triangle adjacent to such a boundary arc (Figure 2.2.4), and suppose a transformation of the normalized triangle is required which maps the node at (0.5,0) onto the point where the perpendicular bisector of the adjacent edge intersects the boundary. The coordinates of the new midpoint node are (Figure 2.2.4):

$$x_6 = 0.5(x_1+x_2) + \beta(y_2-y_1)/d$$

$$y_6 = 0.5(y_1+y_2) - \beta(x_2-x_1)/d$$

where β is the distance that node 6 has been displaced from its original position, and d is the length of the edge connecting nodes 1 and 2.

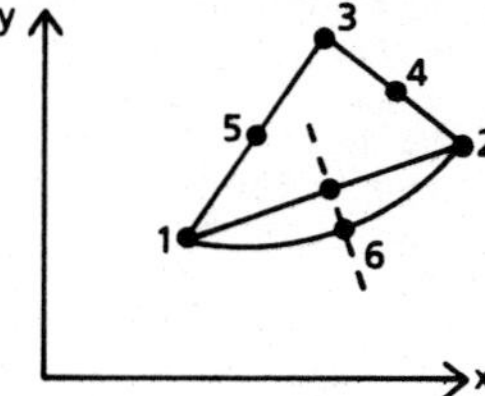

Figure 2.2.4

Then the two transformations:

(2.2.3)
$$\xi' = \xi + (4\xi_6'-2)\xi(1-\xi-\eta)$$
$$\eta' = \eta + 4\eta_6'\xi(1-\xi-\eta)$$

followed by:

(2.2.4)
$$x = (x_2-x_1)\xi' + (x_3-x_1)\eta' + x_1$$
$$y = (y_2-y_1)\xi' + (y_3-y_1)\eta' + y_1$$

where

$$\xi_6' = 0.5 + \beta\{(x_3-x_1)(x_2-x_1) + (y_3-y_1)(y_2-y_1)\}/(d\Delta)$$

$$\eta_6' = -\beta d/\Delta$$

$$\Delta = (y_3-y_1)(x_2-x_1) - (x_3-x_1)(y_2-y_1) = 2*(\text{triangle area})$$

carry the six nodes of the normalized triangle in Figure 2.2.2 onto the six nodes of Figure 2.2.4. The first transformation takes $(0.5,0)$ onto (ξ_6',η_6') and leaves the other five nodes untouched. The second is the same as transformation (2.2.2) and maps (ξ_6',η_6') onto (x_6,y_6) and transforms the other five in the same way as (2.2.2) did previously.

The two together, $x = x(\xi,\eta)$, $y = y(\xi,\eta)$ define what is called an "isoparametric" transformation because $x(\xi,\eta)$ and $y(\xi,\eta)$ have the same form (in this case quadratic polynomials) as the approximating functions themselves. This mapping carries the base of the normalized triangle onto a curve whose parametric equations are quadratic polynomials of the parameter, ξ. The curved triangle in Figure 2.2.4, then, interpolates the boundary with a quadratic rather than a linear fit, and should be expected to give better results than a straight triangle.

Now the basis functions for the approximation (2.1.2) can be defined. They are simply the basis functions ψ_i in (2.2.1) transformed from the normalized to the general triangle. That is, if $\phi_k(x,y)$ is the continuous piecewise quadratic which takes the value 1 at the node with "global" number k, and zero at all other nodes, then in any triangle which contains node k:

$$\phi_k(x,y) = \psi_i(\xi(x,y),\eta(x,y))$$

where i is the "local" node number ($1\leq i\leq 6$) of the node where ϕ_k is 1, and $\xi(x,y),\eta(x,y)$ are found by inverting the transformations (2.2.3-4). In any triangle which does not contain node k, $\phi_k = 0$ identically. The inversion is not easily done, but fortunately it will not need to be done explicitly. Notice that $\phi_k(x,y)$ will not be a polynomial in a curved triangle.

Now the nodes are numbered in such a manner that the nodes on ∂R_1 are given negative numbers $-1,-2...-N1$ and the remaining nodes are given positive numbers $1,2,3...N$. Then ϕ_0 in (2.1.2) can be defined:

$$\phi_0(x,y) = \sum_{i=-1}^{-N1} FB(x_i,y_i)\ \phi_i(x,y)$$

where (x_i,y_i) represents node number i. It is easily verified that $\phi_0 = FB$ (at least at all the nodes on ∂R_1) and that $\phi_1...\phi_N$ are each zero at all these nodes, so that they are identically zero along ∂R_1 (in curved triangles, they are only zero along the quadratic approximation to ∂R_1). This completes the selection of the basis functions needed to define the approximation (2.1.2). Any function in S_2 which satisfies the boundary condition on ∂R_1 can now be expressed:

$$u(x,y) = \phi_0(x,y) + \sum_{j=1}^{N} a_j\phi_j(x,y)$$

where a_j is the value of u at node number j.

The cubic polynomial functions used by PDE/PROTRAN have the form:

$$s(x,y) = a_1+a_2x+a_3y+a_4x^2+a_5xy+a_6y^2+a_7x^3+a_8x^2y+a_9xy^2+a_{10}y^3$$

in each (straight) triangle, and thus 10 nodes per triangle are required to specify the cubic polynomial. The placement of these nodes, in the normalized triangle, is shown in Figure 2.2.5.

The rationale behind the choice of r, as opposed to the obvious choice r=1/3, has to do with convenience in calculating boundary integrals, and will be discussed in Section 2.3.

The linear transformation (2.2.2) will suffice to transform Figure 2.2.5 onto a straight triangle, but for a curved triangle, an isoparametric mapping $x=x(\xi,\eta)$, $y=y(\xi,\eta)$ will be required, where x and y are cubic functions of ξ and η. When the

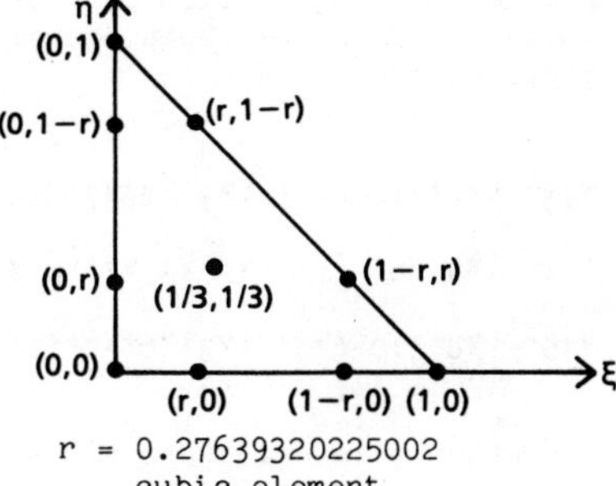

r = 0.27639320225002
cubic element
Figure 2.2.5

triangle is curved, the image of the nodes at (r,0) and (1-r,0) are displaced out perpendicular to the base until they hit the boundary arc (Figure 2.2.6). The image of the node at (1/3,1/3) can be chosen somewhat arbitrarily, and Figure 2.2.6 illustrates the PDE/PROTRAN choice. If v_i represents the position of node number i after the mapping, then:

$$v_{10}= (v_2+v_3+v_5+v_6+v_8+v_9)/6$$

The images of the remaining seven nodes are the same as under the linear transformation. Clearly, a cubic transformation of the form $x=x(\xi,\eta)$, $y=y(\xi,\eta)$ can be constructed

Figure 2.2.6

to map the nodes of Figure 2.2.5 onto any 10 points desired, because, as already claimed, a cubic $(x(\xi,\eta))$ can be found which takes on any 10 assigned values (x_i) at the nodes, and similarly for $y(\xi,\eta)$.

The quartic element has 15 nodes, one for each of the 15 coefficients in a full quartic polynomial, located as shown in Figure 2.2.7.

Again, an isoparametric (quartic)
mapping $x=x(\xi,\eta)$, $y=y(\xi,\eta)$ can be
constructed which maps the nodes
at (s,0), (0.5,0) and (1-s,0) onto
the points where the perpendiculars
to the base intersect the boundary
arc. The images chosen for the
three interior nodes are again
somewhat arbitrary, and Figure
2.2.8 illustrates the choices
made by PDE/PROTRAN. If $\mathbf{v_i}$
represents the position of node
number i after the isoparametric
mapping, then:

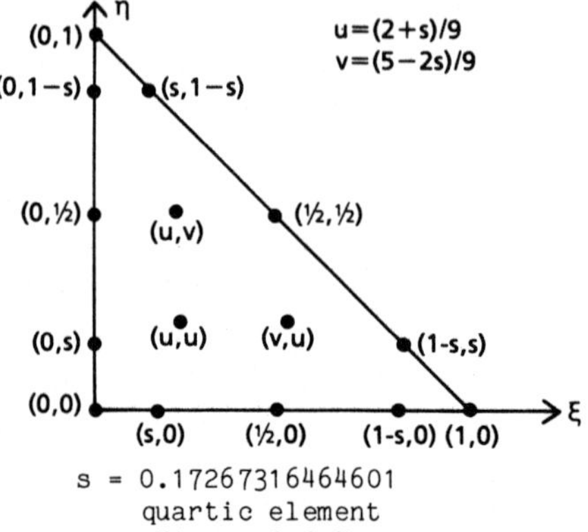

s = 0.172673164646 01
quartic element
Figure 2.2.7

$$\mathbf{v_{13}}=2(\mathbf{v_2}+\mathbf{v_{12}})/9 + (\mathbf{v_3}+\mathbf{v_4}+\mathbf{v_7}+\mathbf{v_{10}}+\mathbf{v_{11}})/9$$

$$\mathbf{v_{14}}=2(\mathbf{v_4}+\mathbf{v_6})/9 + (\mathbf{v_2}+\mathbf{v_3}+\mathbf{v_7}+\mathbf{v_8}+\mathbf{v_{11}})/9$$

$$\mathbf{v_{15}}=2(\mathbf{v_8}+\mathbf{v_{10}})/9 + (\mathbf{v_3}+\mathbf{v_6}+\mathbf{v_7}+\mathbf{v_{11}}+\mathbf{v_{12}})/9$$

The other nine nodes are not moved
when the boundary is curved. Note
that if the triangle has no curved
edge, the above formulas for the
placement of nodes 13,14 and 15
agree with Figure 2.2.7.

The basis functions in the
normalized triangle can be
generated once and for all, and
as in the quadratic case, the
basis functions in the general
triangle are obtained from these
through the isoparametric transformations.

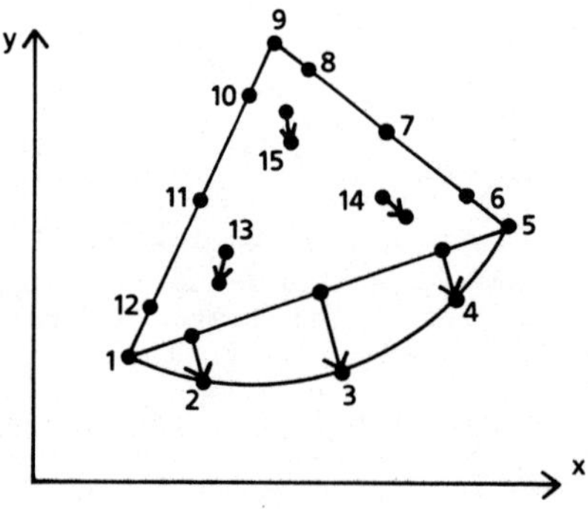

Figure 2.2.8

For all three elements, the basis functions are continuous, but their
derivatives are not continuous across element boundaries. Fortunately, this
is all that is required for the Galerkin method. The integrals in the
Galerkin equations (2.1.4) involve only first derivatives of the basis
functions, not second derivatives, and these are discontinuous but finite.

In the discussion in Section 2.1, it was concluded that if it can be shown
that there exists any member of the approximating set which is close to the
solution u of the PDE (assume the number of PDEs is 1) it may safely be
assumed that the Galerkin solution is comparably close to this true solution.
So, let $u_T(x,y)$ be the piecewise polynomial which, within each triangle, is
the Taylor polynomial interpolant to u, at the midpoint, of degree 2. Then
the error at a given point (x,y) within a (straight) triangle T_k is given by
((x_M,y_M) is the midpoint):

$$u(x,y)-u_T(x,y) = u_{xxx}(x-x_M)^3/6 + u_{xxy}(x-x_M)^2(y-y_M)/2$$

$$+ u_{yyy}(y-y_M)^3/6 + u_{xyy}(x-x_M)(y-y_M)^2/2$$

where the derivatives are evaluated at some point on the line between (x,y) and the midpoint. Thus if $h_k=$ longest side of triangle k, and

$$D_k^n u = \max_{i+j=n} \; \max_{T_k} \; | \partial^n u(x,y)/\partial x^i \partial y^j |$$

then

$$\max_{T_k} \; | u(x,y)-u_T(x,y) | \;\; \leq \;\; 4/3 \; h_k^3 D_k^3 u$$

or

$$\max_{R} \; | u(x,y)-u_T(x,y) | \;\; \leq \;\; 4/3 \; \max_k \; h_k^3 D_k^3 u = 4/3 \; M$$

where $M = \max\limits_{k} \; h_k^3 D_k^3 u$

Unfortunately, u_T is not a member of the approximating space, because it is discontinuous across element boundaries. Nevertheless, a member u_I of that space can be defined by specifying that u_I be equal at each node to the average of the values of u_T at that node. Now since u_T never differs by more than $4/3$ M from u at any point in R, the different values of u_T at any given node cannot vary by more than twice that amount, and thus cannot differ by more than $8/3$ M from their average, u_I. Thus, within triangle T_k:

$$| u_I(x_i,y_i)-u_T(x_i,y_i) | \;\; \leq \;\; 8/3 \; M \qquad \text{at each node } (x_i,y_i)$$

If this difference is called $d(x,y)=u_I(x,y)-u_T(x,y)$, d is a piecewise quadratic polynomial and so from (2.2.1):

$$d(x,y) = \sum_{i=1}^{6} d(x_i,y_i) \; \psi_i(\xi(x,y),\eta(x,y))$$

Therefore at any point in the triangle, since $|\psi_i|\leq 1$ throughout the triangle:

$$| u_I(x,y)-u_T(x,y) | = |d(x,y)| \leq \sum |d(x_i,y_i)| \leq 6*8/3 \; M$$

Finally,

$$\max_{R} |u_I(x,y)-u(x,y)| \leq \max_{R} |u_I(x,y)-u_T(x,y)| + \max_{R} |u_T(x,y)-u(x,y)|$$

$$\leq 52/3 \; \max_k \; h_k^3 D_k^3 u$$

Almost identical arguments show that for any Lagrangian space of piecewise polynomials of degree n, including the PDE/PROTRAN cubic and quartic spaces, there exists a member of that space such that:

$$(2.2.5) \qquad \max_R |u_I(x,y)-u(x,y)| \leq K_n \max_k h_k^{n+1} D_k^{n+1} u$$

(For other spaces, it may not be true that $|\psi_i| \leq 1$ throughout the triangle, but they will be bounded by some constant, provided the nodes are chosen so that the basis functions are uniquely defined.)

Thus if $h = \max h_k$, and if the (n+1)st derivatives of the PDE solution are all bounded in R, the error is of order $O(h^{n+1})$. Strang and Fix {1, Section 4.4} show that with isoparametric transformations of the type used here, under reasonable assumptions, the fact that the boundary is curved does not reduce the order of the error.

It is shown in Problem 2.13 that the interpolation error in the "H-norm" is of order $O(h^n)$, and thus the Galerkin error in this norm is guaranteed to be of order $O(h^n)$ also, for problems derived from energy minimization.

2.3 Numerical Integration

Now whatever method is used to solve the Galerkin equations (2.1.4), it will
certainly be necessary to be able to evaluate the integrals shown there, for
given values of the unknowns a_j. Further, since a Newton-Raphson iteration
(Section 3.1) will be used to solve these equations, it will be necessary to
evaluate the integrals in the formula (2.1.9) for the elements of the
Jacobian matrix. Since in general it will obviously not be possible to do
the required integration analytically, numerical integration is required.

The integrals in these two expressions are "assembled" element by element.
That is, the contributions to all the nonlinear functions f_k and Jacobian
elements J_{kj} made by one triangle are finished before any integrals over the
next triangle are calculated. This is much more efficient than trying to
calculate complete integrals (over R or ∂R_2) at once, since once the basis
functions and isoparametric transformations have been calculated for a given
triangle, there is no sense in coming back later and having to recalculate
them.

A triangle will make contributions only to those functions f_k such that node
number k is on the triangle, since otherwise ϕ_k is zero at all nodes in the
triangle and thus identically zero throughout, and thus the integrals in
(2.1.4) over this triangle vanish. Similarly, it will make contributions
only to those Jacobian elements (really m by m matrices) J_{kj} such that both
nodes j and k are on the triangle, since otherwise at least one of ϕ_j or ϕ_k
vanishes identically in the triangle (see 2.1.9).

From this it is clear that J_{kj} is zero unless nodes j and k have at least
one triangle in common. Thus for large problems most of the Jacobian
elements are zero (the matrix is "sparse"). This is, in fact, a primary
reason that in the finite element method, the functions ϕ_k are chosen to be
piecewise polynomials.

Now assuming that j and k belong to triangle T_ℓ, how are the contributions
of T_ℓ to f_k and J_{kj} calculated? First, the isoparametric (or linear, if T_ℓ
is not curved) transformation is used to transform the integrals over T_ℓ to
integrals over the normalized triangle. The area integrals in (2.1.4)
become:

$$(2.3.1) \qquad \iint_{T_\ell} (F\phi_k - A(\phi_k)_x - B(\phi_k)_y)\ dxdy =$$

$$\int_0^1 \int_0^{1-\xi} \{F(x(\xi,\eta),y(\xi,\eta),\mathbf{u_G},(\mathbf{u_G})_x,(\mathbf{u_G})_y)\ \phi_k(x(\xi,\eta),y(\xi,\eta))$$

$$-A(x(\xi,\eta),y(\xi,\eta),\mathbf{u_G},(\mathbf{u_G})_x,(\mathbf{u_G})_y)\ ((\phi_k)_\xi \xi_x + (\phi_k)_\eta \eta_x)$$

$$-B(x(\xi,\eta),y(\xi,\eta),\mathbf{u_G},(\mathbf{u_G})_x,(\mathbf{u_G})_y)\ ((\phi_k)_\xi \xi_y + (\phi_k)_\eta \eta_y)\} \ (x_\xi y_\eta - y_\xi x_\eta)\ d\eta d\xi$$

The partial derivatives $x_\xi, x_\eta, y_\xi, y_\eta$ are easily calculated, since x and y are
known explicitly as polynomial functions of ξ and η (for quadratics, formulas
2.2.3-4). The functions $\xi_x, \xi_y, \eta_x, \eta_y$ on the other hand, are calculated from:

$$\begin{bmatrix} \xi_x & \xi_y \\ \eta_x & \eta_y \end{bmatrix} = \begin{bmatrix} x_\xi & x_\eta \\ y_\xi & y_\eta \end{bmatrix}^{-1}$$

Now $x_\xi, x_\eta, y_\xi, y_\eta, \xi_x, \xi_y, \eta_x, \eta_y$ are all known explicitly as functions of ξ and η.

The basis functions and their derivatives with respect to ξ and η are also known explicitly, as functions of ξ and η. For,

$$\phi_k(x(\xi,\eta),y(\xi,\eta)) = \psi_i(\xi,\eta)$$

$$(\phi_k)_\xi(x(\xi,\eta),y(\xi,\eta)) = (\psi_i)_\xi(\xi,\eta)$$

and similarly for $(\phi_k)_\eta$, where ψ_i is the normalized triangle basis function which is 1 at the node with global number k. These ψ_i can be calculated once and for all, as can $(\psi_i)_\xi, (\psi_i)_\eta$. (For quadratics the ψ_i are given by 2.2.1).

This takes care of everything in (2.3.1) except $\mathbf{u_G}$ and its derivatives. But these are readily calculated also, since in triangle T_ℓ:

$$\mathbf{u_G}(x,y) = \sum_{i=-1}^{-N1} \mathbf{FB}(x_i,y_i)\phi_i(x,y) + \sum_{i=1}^{N} \mathbf{a_i}\phi_i(x,y) = \sum_{j=1}^{N_0} \mathbf{A_j}\psi_j(\xi(x,y),\eta(x,y))$$

where N_0 is the number of nodes on T_ℓ (6,10 or 15) and $\mathbf{A_j}$ corresponds to either a value of $\mathbf{FB}$ (if on ∂R_1) or a value of $\mathbf{a_i}$ (if not), since $\phi_i(x,y)$ is zero unless node i lies on triangle T_ℓ. So,

$$(\mathbf{u_G})_x = \sum \mathbf{A_j}((\psi_j)_\xi \xi_x + (\psi_j)_\eta \eta_x)$$

and $(\mathbf{u_G})_y$ is likewise calculated.

Now all the integrands in (2.3.1) can be calculated for any given value of (ξ,η), so the problem reduces to calculating:

$$\int_0^1 \int_0^{1-\xi} G(\xi,\eta)\, d\eta d\xi$$

where G is a known function.

The area integrals in the formula (2.1.9) for the Jacobian components have the same type of terms, and so its integrands can also be evaluated for arbitrary (ξ,η), using the techniques above.

A numerical scheme for integration over the normalized triangle consists of a formula of the form:

$$\int_0^1 \int_0^{1-\xi} G(\xi,\eta)\, d\eta d\xi = \sum_{i=1}^{M} w_i G(\xi_i,\eta_i)$$

The integration schemes {1, Table 4.1} used by PDE/PROTRAN for the quadratic, cubic and quartic elements are summarized in Table 2.3.1.

Table 2.3.1

i	ξ_i	η_i	w_i
QUADRATICS - EXACT FOR POLYNOMIALS OF DEGREE 2			
1	0.16666666666667	0.66666666666667	0.16666666666667
2	0.66666666666667	0.16666666666667	0.16666666666667
3	0.16666666666667	0.16666666666667	0.16666666666667
CUBICS - EXACT FOR POLYNOMIALS OF DEGREE 5			
1	0.33333333333333	0.33333333333333	0.11250000000000
2	0.10128650732346	0.10128650732346	0.06296959027241
3	0.79742698535309	0.10128650732346	0.06296959027241
4	0.10128650732346	0.79742698535309	0.06296959027241
5	0.47014206410511	0.47014206410511	0.06619707639425
6	0.05971587178977	0.47014206410511	0.06619707639425
7	0.47014206410511	0.05971587178977	0.06619707639425
QUARTICS - EXACT FOR POLYNOMIALS OF DEGREE 6			
1	0.87382197101700	0.06308901449150	0.02542245318510
2	0.06308901449150	0.87382197101700	0.02542245318510
3	0.06308901449150	0.06308901449150	0.02542245318510
4	0.50142650965818	0.24928674517091	0.05839313786319
5	0.24928674517091	0.50142650965818	0.05839313786319
6	0.24928674517091	0.24928674517091	0.05839313786319
7	0.63650249912140	0.31035245103379	0.04142553780919
8	0.63650249912140	0.05314504984482	0.04142553780919
9	0.31035245103379	0.63650249912140	0.04142553780919
10	0.31035245103379	0.05314504984482	0.04142553780919
11	0.05314504984482	0.31035245103379	0.04142553780919
12	0.05314504984482	0.63650249912140	0.04142553780919

Now consider the boundary integrals in (2.1.4) and (2.1.9). If a triangle is adjacent to the boundary, it is the base of the normalized triangle which is always mapped onto the boundary arc, curved or not. (A triangle is allowed to have at most one edge adjacent to the boundary.) So the isoparametric (or linear, for straight triangles) transformation can be used to reduce an integral over the boundary of triangle T_ℓ to an integral over the base of the normalized triangle:

$$\int_{\partial R_2 \cap T_\ell} \mathbf{GB}\, \phi_k\, ds =$$

$$\int_0^1 \mathbf{GB}(x(\xi,0),y(\xi,0),\mathbf{u_G})\phi_k(x(\xi,0),y(\xi,0))(x_\xi^2+y_\xi^2)^{0.5}\, d\xi$$

Here the fact that $ds = (dx^2+dy^2)^{0.5} = (x_\xi^2+y_\xi^2)^{0.5}\, d\xi$ has been used.

As in the area integral case, all the integrands are readily evaluated for any ξ. The numerical integration schemes of the form:

$$\int_0^1 G(\xi) \, d\xi = \sum_{i=1}^{M} w_i G(\xi_i)$$

used by PDE/PROTRAN are shown in Table 2.3.2 below.

Table 2.3.2

i	ξ_i	w_i
QUADRATICS	- EXACT FOR POLYNOMIALS OF DEGREE 3	
1	0.0	0.16666666666667
2	0.50000000000000	0.66666666666667
3	1.00000000000000	0.16666666666667
CUBICS	- EXACT FOR POLYNOMIALS OF DEGREE 5	
1	0.0	0.08333333333333
2	0.27639320225002	0.41666666666667
3	0.72360679774998	0.41666666666667
4	1.00000000000000	0.08333333333333
QUARTICS	- EXACT FOR POLYNOMIALS OF DEGREE 7	
1	0.0	0.05000000000000
2	0.17267316464601	0.27222222222222
3	0.50000000000000	0.35555555555556
4	0.82732683535399	0.27222222222222
5	1.00000000000000	0.05000000000000

In each case, the integration points are chosen to maximize the degree of accuracy, with the restriction that 0 and 1 be included, and coincide with the nodal points along the base. In fact, this is the basis for the specified positioning of the nodes in the cubic and quartic elements. This simplifies somewhat the calculation of the boundary integrals, especially those for the Jacobian, since by (2.1.9) the boundary integral contribution to J_{kj} is zero at all nodes, and therefore all integration points, unless j=k.

The area integrals are not calculated exactly, even for a simple constant coefficient problem such as $u_{xx}+u_{yy}-u = 0$. However, Strang and Fix {1, Section 4.3} show that the numerical integration schemes must be exact for polynomials of degree 2n-2, where n is the degree of the piecewise polynomial space, in order to avoid decreasing the overall order of accuracy (exponent of h) of the solution. This criterion is met by all three elements, barely.

It should be mentioned here that the partial derivatives F.U, F.UX ... appearing in (2.1.9) are calculated by PDE/PROTRAN from the user supplied functions **F,A,B** and **GB**, using one-sided differences with very large step sizes. This ensures minimal roundoff problems when the function in question is a linear function of the argument, but will generally produce garbage if it is non-linear. In the latter case the user is required to supply the partial derivative explicitly. While in theory this could mean supplying $10m^2$ partial derivatives, in practice only a very few (none if the PDE is linear) need to be supplied. It is felt that this price is worth paying, since taking a small step size can lead to disastrous results, quite frequently.

The choice of integration scheme is of particular importance for the penalty function formulation of the incompressible fluid flow problem (Section 1.3). When the penalty parameter, α, is large, as it must be, the continuity equation $u_x + v_y = 0$ is almost exactly enforced at each integration point. This uses up M degrees of freedom in each triangle, where M is the number of integration points. For the quadratic, cubic and quartic elements there are an average of 2, 4.5 and 8, respectively, nodes per triangle and thus 4, 9 and 16 degrees of freedom available (there are two unknowns per node). M must be less than this number so that there will be some degrees of freedom left for the rest of the fluid flow equations. Since M=3, 7 and 12, respectively, this criterion is satisfied for all three elements. Actually, in straight triangles the continuity equation may be imposed identically throughout the triangle while using up only 3, 6 and 10, respectively, degrees of freedom (Problem 2.6), so even if more integration points are used this would pose a problem only in curved triangles.

Clearly, there are many details regarding efficient assembly of the function vector and Jacobian which cannot all be discussed here. However, the preceding at least outlines the mathematical considerations involved in computing (2.1.4) and (2.1.9).

2.4 Triangulation Refinement and Grading

Given a triangulation of R, the preceding sections outline how to calculate
a piecewise polynomial approximation to the PDE solution, using the Galerkin
method. As discussed in Section 2.1, the accuracy of the Galerkin solution
depends simply on how accurately the true solution **can be** approximated by
functions in the approximating space. For positive definite problems, in
fact, it was proved that the Galerkin solution is the most accurate, in a
certain norm, among all members of the space.

Now in order to accurately approximate the solutions of real-world applications, which very frequently vary much more rapidly in some portions of R
than in others, it is clearly necessary to be able to grade (refine locally)
the triangulation. A finite element program which can generate only uniform
triangulations would be useless as a general tool. Fortunately, triangular
elements are ideally suited for local mesh refinement. In this section, a
formula will be developed which will serve as a rule to guide the mesh
grading.

The starting point for the development of this formula is (2.2.5) which may
be stated:

$$(2.4.1) \qquad \min_{u_I \epsilon S_n} ||u_I - u||_\infty \leq K_n \max_k h_k^{n+1} D_k^{n+1} u$$

where S_n is the Lagrangian piecewise polynomial space of degree n (n=2,3 or
4 for the PDE/PROTRAN elements).

A little geometry shows that $h_k^2 \leq 4A_k/\sin \theta_0$, where A_k is the area of triangle
T_k and θ_0 is the smallest angle of any triangle in the triangulation.
Then,

$$\min_{u_I \epsilon S_n} ||u_I - u||_\infty \leq K_n \max_k (4A_k/\sin \theta_0)^{(n+1)/2} D_k^{n+1} u$$

$$= K_n' \max_k \{ (D_k^{n+1} u)^{2/(n+1)} A_k \}^{(n+1)/2}$$

Now suppose, for the moment, that all (n+1)st derivatives of the solution u
are bounded in R, and define:

$$D^n u(x,y) = \max_{i+j=n} | \partial^n u(x,y)/\partial x^i \partial y^j |$$

Then when the triangles are sufficiently small, so that $D^{n+1} u$ is nearly
constant in each, the inequality can be (approximately) expressed:

$$(2.4.2) \qquad \min_{u_I \epsilon S_n} || u_I - u ||_\infty \leq K_n' \max_k \{ \iint_{T_k} (D^{n+1} u)^{2/(n+1)} dxdy \}^{(n+1)/2}$$

Clearly, to minimize the right hand side of (2.4.2) the triangles should be
chosen so that the indicated integral is evenly distributed over the
triangles. (The sum of the element integrals is fixed, so the maximum is
minimized when all element integrals are equal.)

When this is done, (2.4.2) becomes (NT=number of triangles):

$$(2.4.3) \quad \min_{u_I \varepsilon S_n} \ || u_I - u ||_\infty$$

$$\leq K_n' \ (1/NT)^{(n+1)/2} \ \{ \ \iint_R (D^{n+1}u)^{2/(n+1)} dxdy \ \}^{(n+1)/2}$$

Since for uniform triangulations 1/NT is proportional to h^2, the power of NT in the above formula is what would be expected for smooth problems using a uniform triangulation, so it may be considered "optimal order" in NT.

The interesting thing about (2.4.3) is that the integral there is finite for many functions which do not have bounded (n+1)st derivatives. In fact, it is finite for all functions of the form $u=r^\alpha f(x,y)$, where r is distance to a given point, $0 \leq \alpha$ and $f(x,y)$ is smooth. This includes functions which are bounded but have unbounded derivatives, for example, $u = r^{0.5}$.

Of course, it was assumed during the development of (2.4.3) that u has bounded (n+1)st derivatives. Nevertheless, there is overwhelming experimental and theoretical {2,3} evidence to support the conclusion that (2.4.3) holds for essentially any function such that the right hand side is finite, provided this integral is approximately equi-distributed. In fact, Sewell {4} uses the integral form of the Taylor series remainder to prove that if the integral below is finite and equi-distributed, then:

$$(2.4.4) \quad || u_T - u ||_1 \ \leq K_n \ (1/NT)^{(n+1)/2} \ \{ \ \iint (D^{n+1}u)^{2/(n+3)} dxdy \ \}^{(n+3)/2}$$

where u_T is the piecewise Taylor polynomial of degree n which interpolates to u at the midpoint of each triangle (as defined in Section 2.2). The conclusion is unavoidable that, with proper mesh grading, optimal order accuracy can be obtained even when the solution has singularities in its derivatives.

The PDE/PROTRAN user controls the grading of the triangulation by supplying a function, called D3EST(x,y). PDE/PROTRAN attempts to equidistribute the integral of $D3EST(x,y)^{2/3}$. Ideally, comparing this with (2.4.3), D3EST(x,y) should be equal to:

$$(D^{n+1}u)^{3/(n+1)}$$

which reduces to D^3u when quadratic elements are used (hence the name).

Of course, the user cannot usually estimate $D^{n+1}u$ accurately, but he/she nearly always has a pretty good idea of where the solution is most active and simply supplies a D3EST(x,y) which is largest where the triangulation is to be most dense. For some problems, such as those with re-entrant corners, the form of the singularity is known, and this information can be utilized. In any case, a triangulation which is graded using purely heuristic criteria will nearly always perform considerably better than a uniform one.

The mechanics of the refinement process are as follows. The user supplies
an initial triangulation generally consisting of only as many triangles as
required to define the region. Then the triangle with the largest estimated
$\iint D3EST(x,y)^{2/3} dxdy$ is selected for division. It is divided by a line
from the midpoint of its longest side to the opposite vertex. (The longest
side is chosen to prevent the generation of very small angles--recall that
there is a term $1/\sin \theta_0$ in the estimate of the error.) If this side is not
adjacent to the boundary, the other triangle which shares this side is also
divided. Otherwise the triangle edges will not match up and the basis
functions will not be continuous. The integral estimate in each new triangle
is recalculated, and the process begins again, until the number of triangles
reaches the requested number.

A typical graded PDE/PROTRAN triangulation, generated using $D3EST=r^{-2.5}$
where r is the distance to the crack tip, is shown in Figure 2.4.1. This is
the triangulation used in Example I, Section 1.5.

Consider now the following singular problem:

$$\nabla^2 u -0.25(x^2+y^2)^{-3/4} = 0 \quad \text{in R (=first quadrant of unit circle)}$$

with boundary conditions:

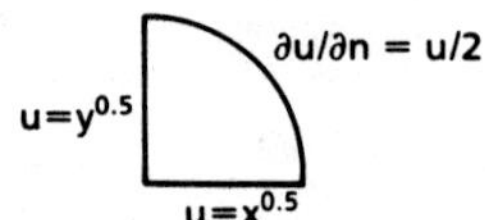

The true solution is $u = r^{0.5}$, which has singularities in its first derivatives. The form of D3EST appropriate to obtain the bound (2.4.3) using
quadratics is $r^{-2.5}$, and the form appropriate for (2.4.4) is $r^{-1.5}$.

Table 2.4.1 shows the PDE/PROTRAN input for one run on this problem, and
Table 2.4.2 and Figure 2.4.2 show the results using the above two grading
schemes plus a uniform refinement sequence (D3EST = 1.0). The measure of
error used is the L_1 norm, which can be calculated automatically using the
PDE/PROTRAN "INTEGRAL" keyword.

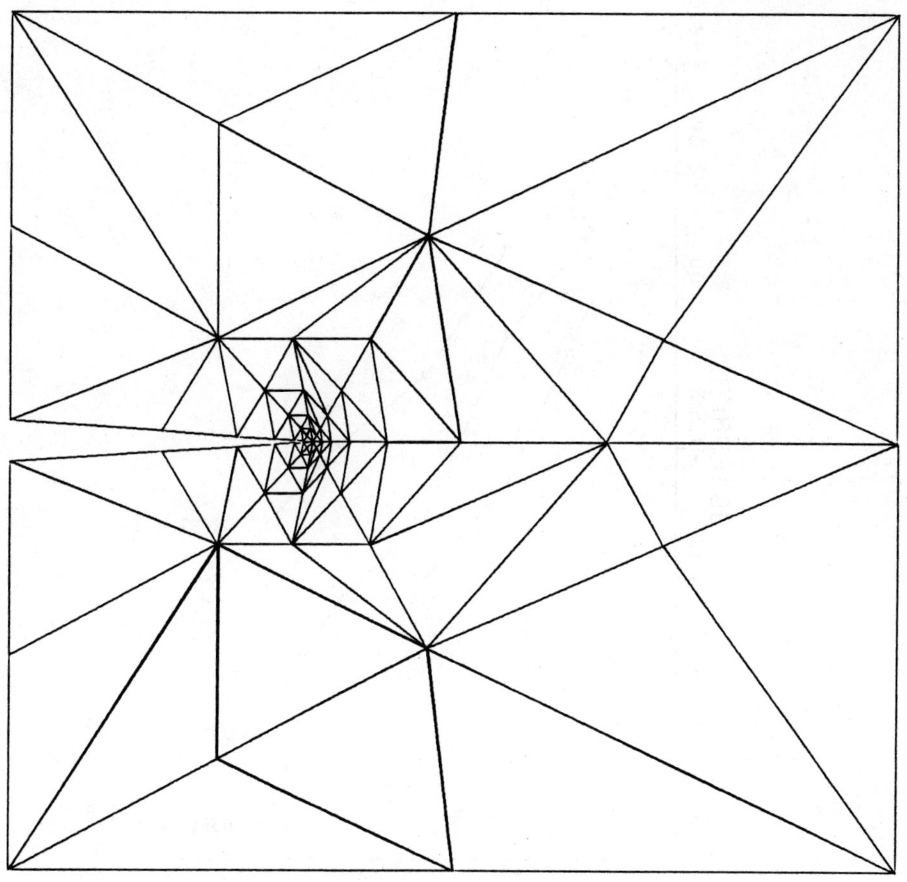

Figure 2.4.1 Graded Triangulation Used for Elastic Crack Problem

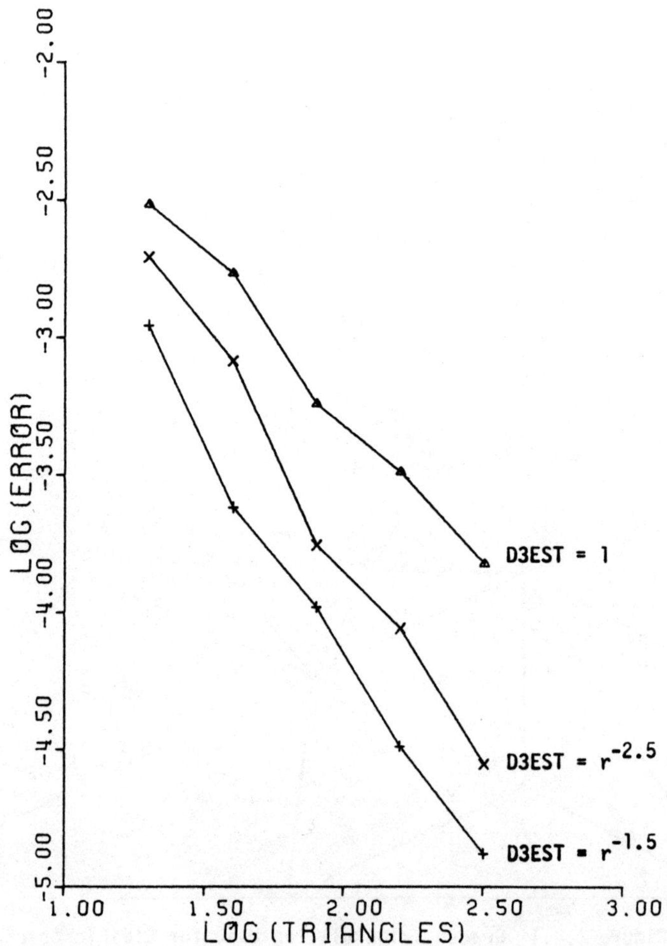

Figure 2.4.2   Errors Using Uniform and Graded Triangulations

Table 2.4.1

```
$       PDE2D
        A = UX
        B = UY
        F = -0.25/(X*X+Y*Y)**0.75
C DEFINE D3EST = R**(-2.5)
        TRIDENSITY = (X*X+Y*Y)**(-1.25)
        SYMMETRIC
        CURVES = (1,DCOS(1.570796327*S),DSIN(1.570796327*S))
        GB = (1,0.5*U)
        FB = (-1,DSQRT(X))  (-2,DSQRT(Y))
        VERTICES = (0,0) (1,0) (0,1) (0.25,0.25)
        TRIANGLES = (1,2,4,-1)  (2,3,4,1)  (3,1,4,-2)
C CALCULATE L1 NORM OF ERROR
        INTEGRAL = DABS(U-(X*X+Y*Y)**0.25)
        NTRIANGLES = 80
        DEGREE = 2
        PRECISION = DOUBLE
        FRONTAL
$       END
```

The average slope of the curve of log(error) versus log(NT) is close to the optimal -1.5 for both graded refinements, but not for the uniform one. The less steep grading did somewhat better, as might be expected since it is the one appropriate for the L_1 error bound.

Table 2.4.2

NT	error (D3EST=1)	error (D3EST=$r^{-1.5}$)	error (D3EST=$r^{-2.5}$)
20	.00303191	.00110298	.00196399
40	.00170448	.00024073	.00082191
80	.00057190	.00010450	.00017531
160	.00032371	.00003260	.00008750
320	.00015068	.00001315	.00002800
average slope of log(error) vs log(NT)	-1.10	-1.57	-1.55

Thus optimal order convergence is observed in spite of the singularity, the curved boundary, and the numerical integration.

43

2.5 **Exercises**   (* = requires use of PDE/PROTRAN)

2.1   Verify formula 2.1.11.

2.2   Verify that transformations (2.2.3) and (2.2.4) map the six nodes in Figure 2.2.2 onto the six nodes of Figure 2.2.4.

2.3   A commonly used triangular element is a linear element with nodes at the three vertices. Give explicit formulas for the the three basis functions in the normalized triangle.

2.4   For the linear element of Problem 2.3, what is the equivalent to the isoparametric transformation 2.2.3?

2.5   Show that the numerical integration scheme with one sample point at (1/3,1/3), with weight 0.5, is exact for polynomials of degree up to one (over the normalized triangle), and thus it is more than accurate enough to use with linear elements.

2.6   Verify the assertion that requiring $u_x + v_y = 0$ identically in a straight triangle uses up only 3,6 and 10 degrees of freedom in the Langrangian elements of degree 2,3 and 4.

2.7   In the discussion of the isoparametric method, it was implicitly assumed that the determinant of the Jacobian of the transformation, $x_\xi y_\eta - y_\xi x_\eta$, does not vanish at any point in the normalized triangle. For the quadratic case, find a condition on β which assures that this does not occur. Interpret your result geometrically.

*2.8   Repeat the problem solved at the end of Section 2.4, using cubic and/or quartic elements. Calculate the experimental order of convergence when D3EST is chosen optimally and compare with the optimal $O(NT^{-2})$ estimate for cubics and $O(NT^{-2.5})$ for quartics. Use high precision.

*2.9   Consider the problem:

$$(Du_x)_x \ + \ (Du_y)_y = 1 \qquad\qquad \text{in } (-1,1) \text{ X } (0,1)$$

with boundary conditions:

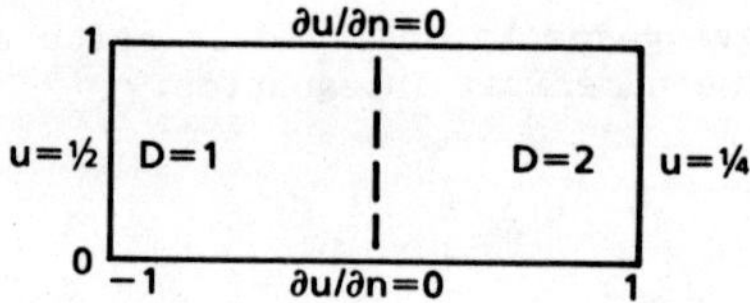

where D = 2 when x is positive and 1 when it is negative.

This type of problem, with discontinuous material properties, frequently arises when composite media are treated.

A finite difference solution requires an explicit interface condition demanding that Du_x be continuous at $x=0$. In the Galerkin finite element equations (2.1.4), however, D appears in undifferentiated form, so it is reasonable to hope that PDE/PROTRAN can handle a discontinuous D directly, without explicit interface conditions. Try it, using the following triangulation:

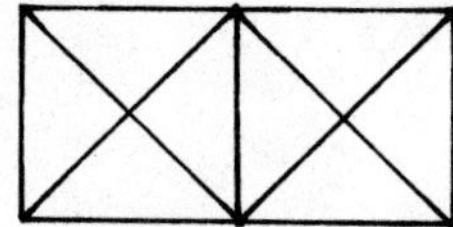

The exact solution is $\quad u = \begin{cases} x^2/4 & x \text{ positive} \\ x^2/2 & x \text{ negative} \end{cases}$

Explain the accuracy (or lack of it) in your solution. If the region were curved, would you expect to achieve the same accuracy?

2.10   Consider the boundary condition:

$$An_x + Bn_y = \beta(u-fb(x,y))$$

When β is a very large number, it appears that this natural boundary condition approximates the essential boundary condition $u = fb(x,y)$. To verify that this "penalty" formulation does work:

a.   Show that when k is a boundary node, the corresponding equation in (2.1.4) reduces to $X + \beta(a_k-fb(x_k,y_k)) = 0$ where X is an expression which is independent of β, (x_k,y_k) are the coordinates of node k, and a_k is the coefficient of ϕ_k in (2.1.2), and so $a_k=u_G(x_k,y_k)$. For large β, therefore, this equation becomes $u_G(x_k,y_k) = fb(x_k,y_k)$. (Hint: recall that the boundary integration points are the nodes, so that ϕ_k is zero at all integration points except node k, and all other ϕ_j are zero at this point.)

b.   Show that when k is not a boundary node, the corresponding equation (2.1.4) does not contain β.

Unlike the "penalty method" for imposing the continuity equation in an incompressible fluid flow problem (Section 1.3), where every equation involves the penalty parameter, here the β appears only in those equations corresponding to the boundary conditions. Thus it would appear that choosing β very large is much less likely to lead to ill-conditioning problems.

*2.11   The boundary condition in Problem 2.10 is primarily useful for handling "mixed" boundary conditions, where essential and natural boundary conditions occur on the same boundary segment. For example, the fourth order clamped plate problem can be written as two second order equations (Section 1.1, with D=1, f=1 for simplicity):

$$\nabla^2 u - v = 0$$

$$\nabla^2 v - 1 = 0 \qquad \text{in R}$$

with

$$u = 0$$

$$\partial u/\partial n = 0 \qquad \text{on } \partial R$$

These "mixed" conditions may be written:

$$u_x n_x + u_y n_y = \partial u/\partial n = 0$$

$$v_x n_x + v_y n_y = \partial v/\partial n = \beta u$$

where β is large so that the second boundary condition is almost u=0. With R equal to the unit circle, solve this problem with β=1.E10, 1.E20, -1.E20. The exact solution is $u = (1-x^2-y^2)^2/64$. Verify that the boundary condition u=0 is imposed accurately on the boundary.

2.12   Another common mixed boundary condition frequently arises in solid and fluid mechanics problems, called the "rolling friction" or symmetry condition. The displacement (or velocity) normal to the boundary is zero, and the boundary force (or traction) tangent to the boundary is zero. Show that, for β large, the boundary conditions:

$$\mathbf{GB} = (\beta n_x(u n_x + v n_y) \; , \; \beta n_y(u n_x + v n_y))$$

impose both desired conditions simultaneously. (Hint: recall that $\mathbf{GB}$ is the boundary force (or traction) vector.)

2.13   Show that with u_I defined as in Section 2.2,

$$|| (u_I)_x - u_x ||_\infty \le K_n' \max_k (h_k^n D_k^n u)$$

(Hint: $(u_T)_x$ is the Taylor interpolant to u_x of degree n-1, so $(u_T)_x - u_x$ is easily bounded. Next differentiate:

$$u_I(x,y) - u_T(x,y) = \sum (u_I(x_i,y_i) - u_T(x_i,y_i)) \; \psi_i(\xi(x,y), \eta(x,y))$$

with respect to x and, making appropriate assumptions on the shape of the triangles, bound $(u_I)_x - (u_T)_x$.)

Use this result to show that the interpolation error in the "H-norm" defined in Section 2.1 (assume 2.1.1 is linear) is of order $O(h^n)$, provided u is smooth.

*2.14 Consider the steady state flow of an incompressible fluid at low Reynold's number (equations 1.3.3 with $\rho=0$) through a pipe and past a spherical obstacle, as shown below:

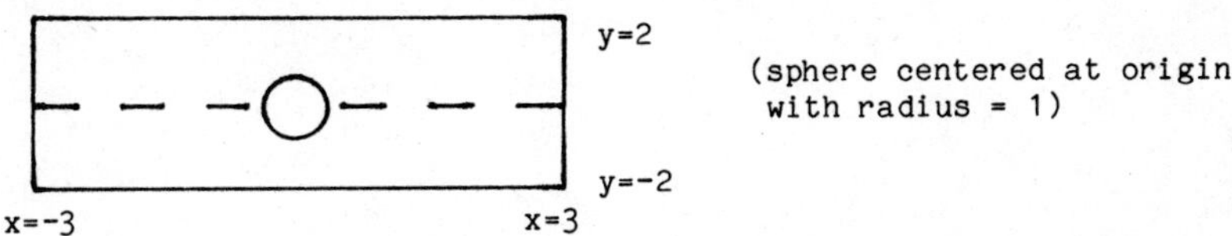

(sphere centered at origin with radius = 1)

Take $\mu=0.1$ and assume that the inlet and outlet velocities are:

$$u = v_0(1-(y/2)^2)$$
$$v = 0$$

where the maximum velocity, v_0, is taken to be 5. Assume no-slip ($u=v=0$) conditions on the rest of the boundary (remember that the sphere is part of the boundary). Notice that y=0 is a line of symmetry for the flow, so that the region R in which the PDEs are solved may be taken to include only the upper half of the above figure, with boundary conditions $v=0$, $\sigma_{12}=GB1=0$ (since $u_y=v_x=0$) imposed along the line of symmetry (cf. Problem 2.12).

a. Use PDE/PROTRAN with a penalty formulation (see Problem 1.2) to solve this problem, and plot the velocity field. Set $\alpha=10^6\mu$ and use high precision. See Figure 2.5.1.

b. Use PDE/PROTRAN with a stream function formulation (1.3.4) to solve this problem, and plot the velocity field. On that part of the boundary where the velocity $(\phi_y,-\phi_x)$ is given, the normal and tangential derivatives of ϕ are known. Along the line of symmetry the tangential derivative of ϕ (v) is zero and the vorticity ($\omega = v_x-u_y$) is also zero. Since the stream function is unique only to within an additive constant, choose $\phi=0$ at some corner and use its tangential derivative to assign values to ϕ along the entire boundary. Problem 2.11 shows how to formulate the boundary conditions when ϕ and $\partial\phi/\partial n$ are given.

Is the penalty or stream function formulation easier to implement on PDE/PROTRAN?

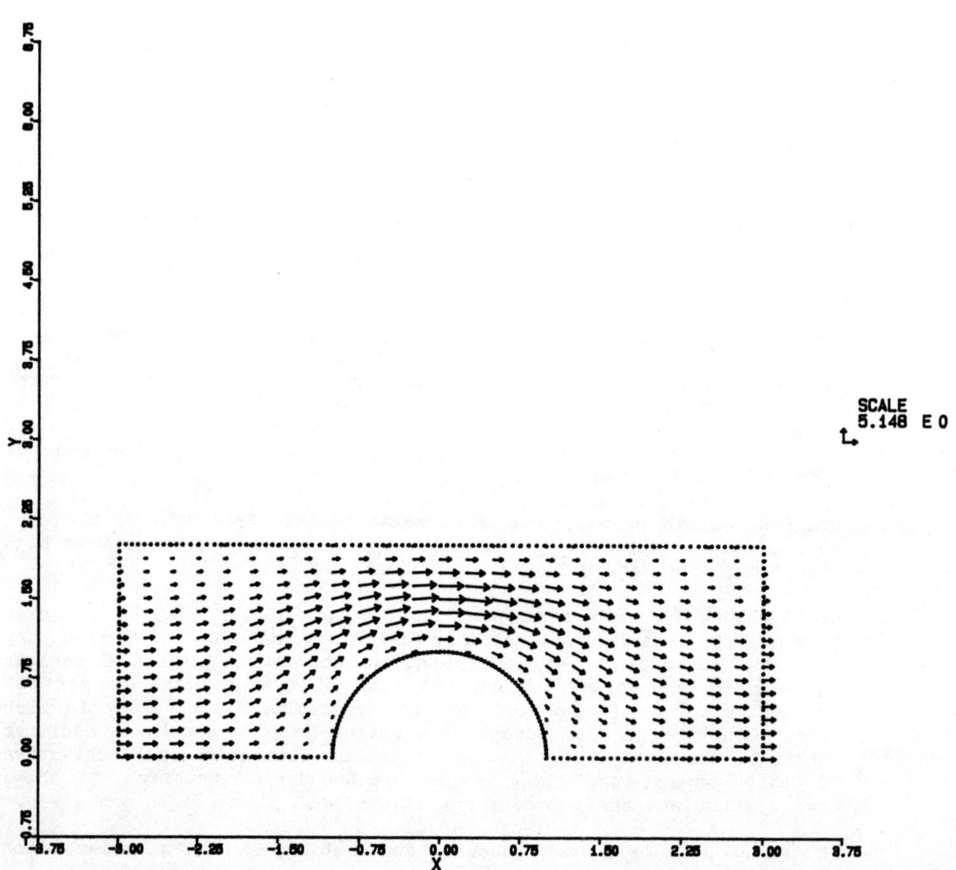

References

1. G. Strang and G. Fix, <u>An Analysis of the Finite Element Method</u>, Englewood Cliffs, N.J., Prentice-Hall, 1973.

2. H. G. Burchard, "Splines (with Optimal Knots) are Better," <u>J. App. Anal. 3</u> (1974), pp309-319.

3. V. Pereyra and G. Sewell, "Mesh Selection for Discrete Solution of Boundary Problems in Ordinary Differential Equations," <u>Numer. Math. 23</u>, (1975) pp261-268.

4. G. Sewell, "Automatic Generation of Triangulations for Piecewise Polynomial Approximation," Ph.D. Thesis, Purdue University (1972).

CHAPTER 3.   ELLIPTIC PROBLEMS - SOLVING THE ALGEBRAIC EQUATIONS

3.1 The Newton-Raphson Method

As outlined in Chapter 2, the finite element approximation to the solution of an elliptic PDE leads to a set of algebraic equations (2.1.4) which are linear or nonlinear as the PDE itself is linear or nonlinear. In either case, this system may be written as:

$$\mathbf{f}(\mathbf{x}) = \mathbf{0}$$

where the number of equations and the number of unknowns is equal to the product of the number of PDEs and the number of "free" nodes (not on ∂R_1).

The basic algorithm used by PDE/PROTRAN is the standard Newton-Raphson iteration, which is a fairly obvious generalization of the one variable Newton-Raphson method:

$$(3.1.1) \qquad \mathbf{x}^{k+1} = \mathbf{x}^k - J(\mathbf{x}^k)^{-1}\mathbf{f}(\mathbf{x}^k)$$

where $J(\mathbf{x}^k)$ is the Jacobian of $\mathbf{f}$ at $\mathbf{x}^k$, that is, the matrix whose ith row is the gradient of f_i.

To study the convergence properties of this iteration, let $\mathbf{x}^*$ be a solution, that is, $\mathbf{f}(\mathbf{x}^*) = \mathbf{0}$. Then:

$$(3.1.2) \qquad J(\mathbf{x}^k) (\mathbf{x}^{k+1}-\mathbf{x}^*) = J(\mathbf{x}^k) (\mathbf{x}^k-\mathbf{x}^*) - \mathbf{f}(\mathbf{x}^k) = E$$

The right hand side is a vector whose ith component is:

$$E_i = \nabla f_i(\mathbf{x}^k)^T(\mathbf{x}^k-\mathbf{x}^*) - f_i(\mathbf{x}^k)$$

According to the multivariate Taylor's series with remainder,

$$0 = f_i(\mathbf{x}^*) = f_i(\mathbf{x}^k) + \nabla f_i(\mathbf{x}^k)^T(\mathbf{x}^*-\mathbf{x}^k) + 0.5(\mathbf{x}^*-\mathbf{x}^k)^T H_i(\xi)(\mathbf{x}^*-\mathbf{x}^k)$$

where $H_i(\xi)$ is the Hessian matrix of f_i ($H_{pq} = \partial^2 f_i/\partial x_p \partial x_q$) evaluated at some point ξ on the line between $\mathbf{x}^*$ and $\mathbf{x}^k$. And so the ith component of the right hand side in (3.1.2) is:

$$(3.1.3) \qquad E_i = 0.5(\mathbf{x}^*-\mathbf{x}^k)^T H_i(\xi) (\mathbf{x}^*-\mathbf{x}^k)$$

If H_{max} represents the maximum of the infinity norm of this Hessian matrix in a certain neighborhood N_1 of $\mathbf{x}^*$, then from (3.1.3) it is deduced that if $\mathbf{x}^k$ is in this neighborhood:

$$| E_i | \leq 0.5 || \mathbf{x}^*-\mathbf{x}^k ||_\infty^2 H_{max}$$

If it is further assumed that $J(\mathbf{x}^*)$ is nonsingular, there must exist a neighborhood N_2 of $\mathbf{x}^*$ within which $||J(\mathbf{x})^{-1}||_\infty$ is bounded, say less than J_{max}. Then within $N_1 \cap N_2$:

$$x^{k+1} - x^* = J(x^k)^{-1}E$$

$$|| x^{k+1} - x^* ||_\infty \le J_{max}||E||_\infty$$

$$\le 0.5 J_{max}H_{max} || x^k - x^* ||_\infty^2$$

Finally, inside the neighborhood N_3 defined by:

$$|| x - x^* ||_\infty < 1/(J_{max}H_{max})$$

every iteration decreases the error:

$$|| x^{k+1} - x^* ||_\infty \le 0.5|| x^k - x^* ||_\infty$$

Therefore, if x^0 is sufficiently close to x^* (in $N_1 \cap N_2 \cap N_3$), the Newton-Raphson method converges, and it converges quadratically. That is, the error on each new iteration is proportional to the square of the error on the previous iteration.

If the system of equations is linear, then all second derivatives are zero, and the error is zero after a single iteration. Thus linear systems are not treated differently from nonlinear systems, except that only one iteration is done, giving the exact solution (to within roundoff error) regardless of the initial guess x^0. Actually, it has been observed that for ill-conditioned linear systems, it is sometimes beneficial to do an extra iteration or two of Newton's method to diminish the roundoff error. This is because Newton's iteration for linear systems is essentially equivalent to iterative refinement, as can be seen by expressing $f(x) = Ax-b$ and substituting into (3.1.1):

$$x^{k+1} = x^k - A^{-1}(Ax^k-b)$$

Although the residual Ax^k-b is not calculated in extended precision as is normally required to make iterative refinement useful, it is calculated directly, without any matrix-vector multiplications, and thus more accurately.

In point of fact, it is not feasible to calculate $J(x^k)^{-1}$ in (3.1.1), particularly since the inverse of a sparse matrix may be full. The Newton iteration is actually carried out by solving a linear system:

$$J(x^k)d^k = f(x^k)$$

(3.1.4)

$$x^{k+1} = x^k - d^k$$

Now if the PDE system is only mildly nonlinear, it may not be difficult for the user to choose a starting solution close enough to the true one to assure convergence. For highly nonlinear problems, however, there are two artifacts available to the PDE/PROTRAN user to increase the probability of convergence.

The first is a relative of the "damped" Newton iteration, which replaces (3.1.4) by:

$$\mathbf{x}^{k+1} = \mathbf{x}^k - \alpha\mathbf{d}^k \qquad\qquad 0 < \alpha < 1$$

Define $g(\alpha) = \mathbf{f}(\mathbf{x}^k-\alpha\mathbf{d}^k)^T\mathbf{f}(\mathbf{x}^k-\alpha\mathbf{d}^k)$ so that $g(\alpha)$ represents the sum of squares of the residuals at $\mathbf{x}^k-\alpha\mathbf{d}^k$. Then:

$$g'(\alpha) = -2*\mathbf{f}(\mathbf{x}^k-\alpha\mathbf{d}^k)^T J(\mathbf{x}^k-\alpha\mathbf{d}^k)\mathbf{d}^k$$

and

$$g'(0) = -2*\mathbf{f}(\mathbf{x}^k)^T J(\mathbf{x}^k) \ J(\mathbf{x}^k)^{-1}\mathbf{f}(\mathbf{x}^k) \ = -2*\mathbf{f}(\mathbf{x}^k)^T\mathbf{f}(\mathbf{x}^k)$$

which is negative, unless $\mathbf{x}^k$ is the solution. This means that if α is chosen sufficiently small, the sum of squares of the residuals can be decreased each iteration.

Unfortunately, damped Newton may lead the iteration into a local minimum of $||\mathbf{f}(\mathbf{x})||_2$ which is not a root, where it may become trapped. In practice, a more effective form of damping appears to be:

$$\mathbf{x}^{k+1} = \mathbf{x}^k - D \ \mathbf{d}^k$$

where D is a diagonal matrix with entries:

$$D_{ii} = \min(1 \ , \ 0.3||\mathbf{x}^k||_\infty/|d_i^k| \ )$$

This steplimitation scheme limits the change per iteration in any component to 30% of the norm of $\mathbf{x}^k$ (and thus zero initial values should not be used!).

This is the algorithm used in PDE/PROTRAN, and it eliminates the primary danger of the Newton method (even in the one variable case) which is that an iterate will hit near a singularity of the Jacobian and be sent off on a long trip by the ensuing large stepsize, and never return. Unlike the damped Newton algorithm, however, this iteration can still bounce around to a limited extent and escape a local minimum. Quadratic convergence is still automatically assured once the iteration gets close to the root, since the d_i's will become small and D will reduce to the identity.

The second trick available to the PDE/PROTRAN user is the employment of continuation. The user is encouraged, for highly nonlinear problems, to parameterize the PDE itself with a parameter (say β, although for convenience the time variable T is actually used) such that when $\beta=0$ the PDE is linear (or otherwise easy to solve with the user's initial guess) and such that when $\beta=1$ it reduces to the desired problem. For example, the PDE $\nabla^2u-e^u=0$ might be parameterized as $\nabla^2u-e^{\beta u}=0$. The Newton iteration begins with $\beta=0$ and each iteration β is gradually increased, until when $\beta=1$ several final iterations are done. If the problem is linear when $\beta=0$, any initial values will do.

This trick is analogous to holding a carrot in front of a donkey and moving it a little every time he gets close. It takes advantage of the fact that Newton-Raphson works well when close to the root, but not so well when distant. With the proper parameterization, a very stubborn donkey can be led to the root.

3.2 **The Band Solver**

Whether the algebraic system (2.1.4) is linear or nonlinear, then, a linear
system of the form (3.1.4) must now be solved. Since for large problems the
majority of the computing time will be spent solving this system, efficiency
is of great importance.

It will initially be assumed that a single PDE is being solved, so that
there is one unknown and one equation associated with each "free" node (not
on ∂R_1). For direct methods, based on Gaussian elimination, the numbering
of these free nodes is critical, as that determines the ordering of the
unknowns and equations and thus the location of the non-zero elements of the
Jacobian matrix (2.1.9).

One of the linear equation solvers available with PDE/PROTRAN is a band
solver, so that the node numbering needs to be done in such a manner that
the "half band width", $b = \max_{a_{ij} \neq 0} |i-j|$ is minimized.

It is clear in Figure 3.2.1 that, provided no pivoting is done, Gaussian
elimination can be carried out on a band matrix without disturbing the zero
elements outside the band. Referring to this figure, a multiple of the

pivot row can be added to each row
below it, to eliminate the
subdiagonal elements in the
pivot column, and only those
elements inside the box will
change. It is further clear
from this discussion that the
number of multiplications required
in the elimination stage is about
Nb^2, assuming $b \ll N$. (The back
substitution stage requires much
less work, about Nb multiplications.)

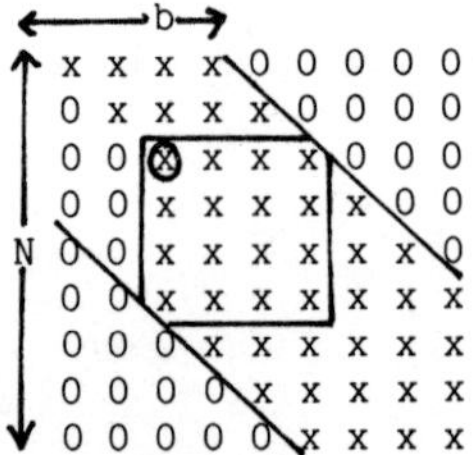

Figure 3.2.1

If pivoting for numerical stability is done, not only can this cause some
initially zero elements above the band to become nonzero ("fill-in") but it
will have an even more destructive effect on the "profile" or "skyline" of
the matrix, and on the frontal method, both discussed later.

Fortunately, as discussed in Section 2.1, for many applications the Jacobian
matrix is positive (or negative) definite, and it is well known that pivoting
is unnecessary for positive (or negative) definite systems.

Although many other applications involve nonsymmetric and/or indefinite
Jacobians, the diagonal elements are never zero initially, and many
PDE/PROTRAN tests, as well as the experiences of other workers, have confirmed
that pivoting is almost never essential for stability. That is, in applications, the probability of a pivot element hitting on or near enough to zero
to cause numerical damage is extremely small.

If a system of m PDEs is being solved, the Jacobian is treated (see 2.1.9) as an N by N matrix whose elements consist of m by m blocks, and all of the preceding discussion still applies. For example, when the ith row is multiplied by $-a_{ji}a_{ii}^{-1}$ and added to the jth row it is understood that a_{ii}^{-1} is the inverse of an m by m matrix. Thus it may be considered that some limited pivoting is done, since this inverse is calculated with pivoting. The importance of this "pivoting" is seen in the following example. The Jacobian matrix resulting from the PDE system:

$$a\nabla^2 u_1 + b\nabla^2 u_2 = f_1(x,y)$$

$$c\nabla^2 u_1 + d\nabla^2 u_2 = f_2(x,y) \qquad \text{in } R$$

with

$$u_1 = u_2 = 0 \qquad \text{on } \partial R$$

will have "elements" $P_{ij}A$, where $A = \begin{bmatrix} a & b \\ c & d \end{bmatrix}$ and the P_{ij} are the elements of an N by N positive definite matrix (see 2.1.9). Provided only that A is nonsingular, it is easy to see that no pivoting will be necessary (other than in the calculation of the inverses of the small pivot matrices). At any stage in the elimination the elements will be $P_{ij}'A$ where P_{ij}' are the elements of the positive definite matrix at the same stage. This, even though the Jacobian may have all diagonal elements equal to zero, if a=d=0.

Returning now to the central problem, which is the ordering of the nodes, it is evident that the objective is to produce a numbering such that nodes which are "neighbors" have numbers close together. Two nodes i and j are called neighbors if they share at least one common triangle, so that the supports of their respective basis functions overlap, and thus $a_{ij} \neq 0$, or at least is not necessarily zero.

The numbering scheme used by PDE/PROTRAN is essentially the reverse Cuthill-McKee algorithm {1}, which undoubtedly is the most widely used. A bandwidth reduction algorithm is then invoked to improve the ordering further.

The Cuthill-McKee algorithm begins with a root node, whose selection will be discussed shortly, which is numbered 1. This node will be the unique member of level 0. Next all neighbors of node 1 are given consecutive numbers 2 through, say, p. These nodes comprise level 1. Next all as-yet-unnumbered neighbors of node 2 are given numbers beginning with p+1, then all neighbors of node 3, and so on until all neighbors of all level 1 nodes are numbered, each new node being assigned the lowest unused number available. All these new nodes are in level 2. This process continues until all nodes are numbered. Most implementations specify the order in which neighbors of a given node are numbered, but this is of marginal importance in large problems.

To illustrate this process, it is convenient to use linear triangular elements (with nodes at the vertices), even though PDE/PROTRAN does not use such elements, because they have the property that each pair of neighboring nodes is connected by an edge. If the higher order elements are considered, a graph in which all nodes which are neighbors are connected by lines can be drawn and all of the following applies. However, the graph will no longer look like the triangulation itself.

Figure 3.2.2 shows a triangulation of linear elements, and the resulting
ordering, with levels indicated.
The level of a node can be seen to
be the length of the shortest path
to the root, where the length of
any edge is taken to be one.

Now i and j are neighbors only if
they belong to the same or adjacent
levels, because if j is connected to
i, then the shortest path from j to
the root is less than or equal to
the shortest path from i to the root plus 1.   Thus level(j) $\leq$ level(i)+1
and similarly, level(i) $\leq$ level(j)+1.   Therefore the half bandwidth, b, is
necessarily less than twice the number of nodes in the most populated level.

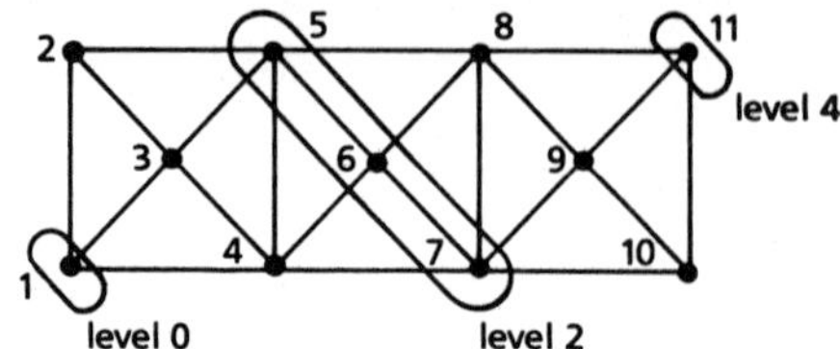

Figure 3.2.2

Geometrically, it seems apparent that the number of nodes in a single level
should be approximately proportional to $N^{0.5}$, provided the refinement is
anisotropic (independent of direction), since a level is a one-dimensional
"front" in a two dimensional graph.   For example, if each triangle in Figure
3.2.2 is divided into four equal triangles,
the result is seen in Figure 3.2.3.
The number of nodes has quadrupled
while the number of levels and the
number of nodes per level have doubled
approximately (exactly in the limit
of infinite refinement of this sort).
In fact, as a triangulation is
refined, the relation b α $N^{0.5}$ is
verified experimentally, and so the
number of multiplications required by
the band solver is of order Nb^2 or N^2.

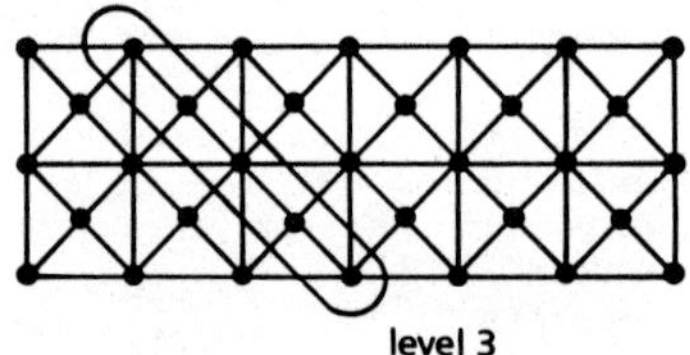

Figure 3.2.3

Since the average number of nodes per level is equal to N divided by the
number of levels, it is reasonable to want to make the number of levels as
large as possible, in hopes of decreasing the bandwidth.   This means choosing
the starting node to be one end of a "diameter"--that is, a pair of points
such that the shortest path between them is as long as possible in the
graph.   In Figure 3.2.2, if the middle node (currently numbered 6) were
chosen as the root, there would be only 3 levels in the graph, rather than
the current 5, so that this would be a poor choice.

A reasonable choice for the root would be any corner of the triangulation.
But a heuristic method to find a better root, used by PDE/PROTRAN, is as
follows.   The free node farthest from the origin is chosen as an initial
root.   The Cuthill-McKee scheme is applied and the last numbered node
becomes the root actually used.   This root usually leads to a smaller
bandwidth than when the original root is used, but its choice is based only
on the intuitive idea that the last node in any Cuthill-McKee ordering is
likely to be near the end of a diameter.

After the Cuthill-McKee ordering has been assigned, a homemade band reduction algorithm is applied to this ordering. The basic idea is to look for those nonzero elements a_{ij} farthest from the diagonal, and try to renumber i or j to make the two closer together. The algorithm is outlined below:

```
1 start
    calculate b_k = max (|i-j|, such that nodes i and j are both on T_k)
    calculate b = max b_k
    if (more than 25% of triangles have b_k=b) stop
    do 2 m=1,mtry
    do 2 k=1,nt
       If (b_k < b) go to 2
       ilow = lowest node number on triangle k
       switch nodes ilow and ilow+m
       recalculate b_k in those triangles which contain node ilow or
               ilow+m, and which have b-m ≤ b_k.
       calculate b = max b_k
       if (b decreased or number of triangles with b_k=b decreased) go to 1
       switch ilow and ilow+m back
       ihigh = highest node number on triangle k
       switch nodes ihigh and ihigh-m
       recalculate b_k in those triangles which contain node ihigh or
               ihigh-m, and which have b-m ≤ b_k.
       calculate b = max b_k
       if (b decreased or number of triangles with b_k=b decreased) go to 1
       switch ihigh and ihigh-m back
2 continue
    stop
    end
```

mtry is set to 4,6 and 8 if quadratic, cubic or quartic elements are used, respectively. nt is the number of triangles.

The algorithm stops when it is unable to decrease b further. The extent of the reduction effected varies greatly, but typically the bandwidth is decreased 10-20%, which is more than enough to make up for the time consumed by the reduction, almost always, and particularly when several systems must be solved on the same triangulation, as for nonlinear or time-dependent problems.

In Figure 3.2.4 a typical nonzero structure for a matrix using the Cuthill--McKee (plus bandwidth reduction) algorithm is shown, with nonzero elements indicated by X's. In each column, all the zeros above the first nonzero element will remain zero throughout the elimination, even if they lie within the band, because a multiple of any row is only added to a lower row, assuming no pivoting. There exist "skyline" or "profile" algorithms which take full advantage of this and store only those elements below the first nonzero and above the diagonal in each column (plus the symmetric elements on the other side of the diagonal--the nonzero structure is always symmetric, even when the matrix is not). Those elements which are initially zero but must be stored by a profile algorithm because they may become nonzero due to fill-in, are indicated in Figure 3.2.4 by the symbol -.

While PDE/PROTRAN stores the entire band, because the savings in storage seems not to be worth the increased awkwardness in the data structure (and the frontal method is available when storage is limited), the fact that not all of the band will fill-in is taken advantage of automatically. Because, if a_{ij} lies above the first nonzero in its column, a_{ji} lies before the first nonzero in its row, and is also protected from fill-in, as is easily verified. So the multiple of row i which should be added to row j is zero, and that addition is skipped. Thus the less fill-in the better, even for a band algorithm. Notice that the fact that no pivoting is done is crucial here.

Now if the Cuthill-McKee ordering is reversed, the result is as shown in Figure 3.2.5. Much less fill-in occurs and the skyline lets in a lot more light. Thus it is clear why the "reverse Cuthill-McKee" ordering is preferred. In actual tests with PDE/PROTRAN, reversing the ordering has cut the computing time by a factor of 10 or more, especially when higher order elements are used.

To see why the reverse ordering improves the profile so drastically, consider the cubic or quartic elements, which have interior nodes. When an interior node receives its number (say p) from the algorithm it is because it is a neighbor to an earlier numbered node (say q), which must belong to the same triangle, since p has neighbors only in that triangle. But PDE/PROTRAN numbers the neighbors of q triangle by triangle, so that all nodes in the triangle will have received their numbers very shortly after p. Thus the highest numbered node in p's triangle will be numbered p+e where e is a very small integer (less than 6, 10 or 15 depending on the element degree). Reversing the ordering means that now the interior node in question is numbered N+1-p and the lowest numbered node in the triangle will be N+1-p-e. Therefore column N+1-p will have its first nonzero entry in row N+1-p-e, that is, very close to the diagonal.

This explains why columns corresponding to interior nodes have a low profile, but Figures 3.2.4-5 result from a quadratic element triangulation, and these elements have no interior nodes. Nevertheless the midpoint nodes are shared by only 2 triangles and a similar, but more complicated, argument can show why they have lower profiles than the vertices, but not as low as interior nodes.

The ordering algorithms of this section are all based on heuristic, or intuitive, arguments. But this is the state of the art, and algorithms guaranteed to produce a minimal bandwidth or profile would be prohibitively time consuming and not at all justified by the further reduction in solution time which might result.

It should be pointed out here that since the innermost DO loop of a band solver adds a multiple of one row to another, of length $O(N^{0.5})$, the band solver can take good advantage of the vector processing capabilities of a vector computer. Since, in addition, PDE/PROTRAN stores the band matrix row-wise, the band solver works efficiently in a virtual memory environment.

Figure 3.2.4 Nonzero Structure Using Forward Cuthill-McKee Ordering

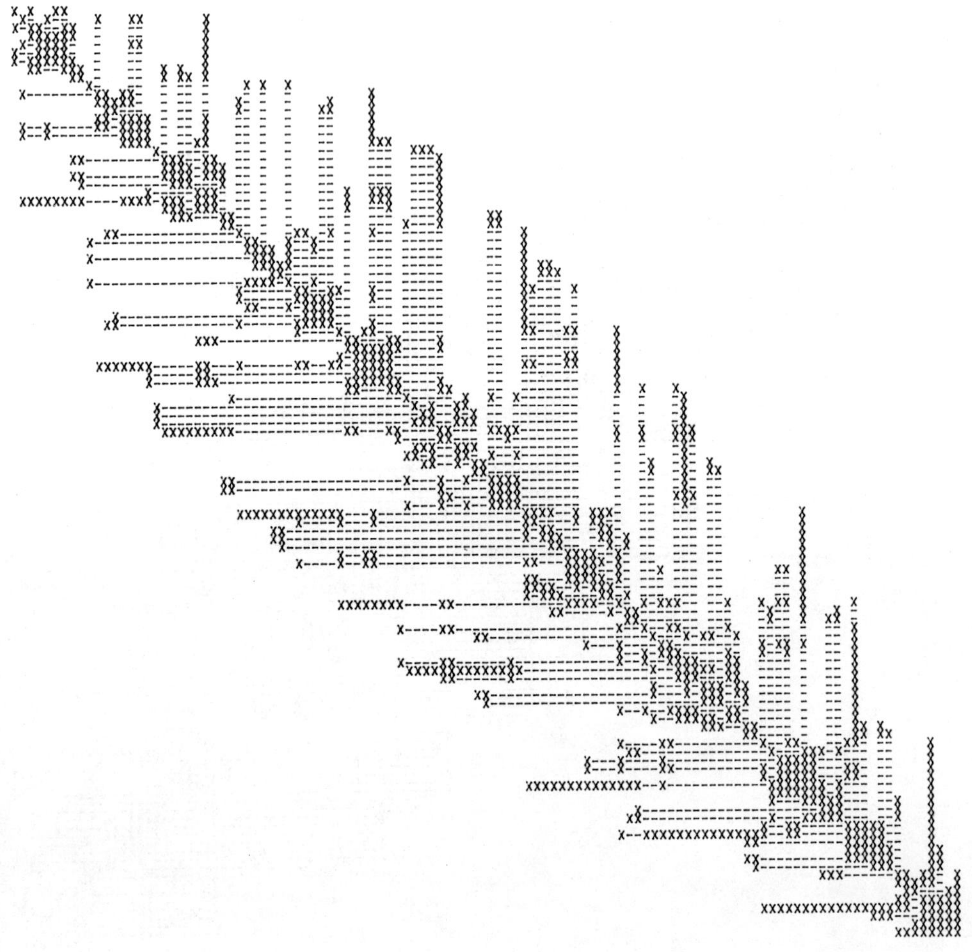

Figure 3.2.5 Nonzero Structure Using Reverse Cuthill-McKee Ordering

60

If the matrix is symmetric, as is often the case, the unfactored portion of
the matrix remains symmetric. To see this, consider what happens to element
a_{ij} (which may be an m by m matrix) when a multiple of the first row is added
to the ith row (see Figure 3.2.6). The new a_{ij} is given by:

$$a_{ij}' = a_{ij} - a_{i1} \, a_{11}^{-1} \, a_{1j}$$

while the new a_{ji} is given by:

$$a_{ji}' = a_{ji} - a_{j1} \, a_{11}^{-1} \, a_{1i}$$

But

$$a_{ji}'^{T} = a_{ji}^{T} - a_{1i}^{T} a_{11}^{-T} a_{j1}^{T} = a_{ij} - a_{i1} a_{11}^{-1} a_{1j}$$

which is just equal to a_{ij}',

so that a_{ij} is still equal to
the transpose of a_{ji}, as
required for a symmetric
matrix. Of course the new
values of a_{i1} and a_{j1} are 0.
The same argument is now applied
to the remaining N-1 by N-1
matrix, and so on. Thus it is
not necessary to store or
compute with the subdiagonal
elements of the matrix, and
PDE/PROTRAN does take advantage
of symmetry in this way. Note that
the absence of pivoting is crucial
again.

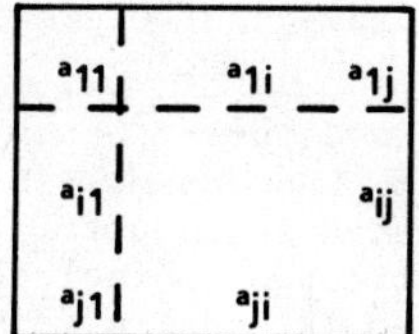

Figure 3.2.6

3.3 The Frontal Solver

The amount of storage required by the band solver is about 2Nb (or Nb for
symmetric matrices) numbers, which is of order $O(N^{1.5})$. For large N, this
can easily exceed the amount of available fast memory. The frontal method
{2} is a technique for organizing the band solver calculations in such a way
as to drastically decrease the amount of in-core storage required, without
increasing the computation time greatly.

First, the triangles are numbered in such a way that if m_k = (minimum node
number on triangle number k), then $m_1, m_2, \ldots$ forms a nondecreasing sequence.
Then by the definition of the half-band width b, the highest numbered node
on triangle k has number $\leq m_k + b$, and so this triangle makes contributions
only to elements a_{ij} with $m_k \leq i \leq m_k + b$, $m_k \leq j \leq m_k + b$, i.e., elements inside the
solid box in Figure 3.3.1. Thus while triangle k is being "assembled" (its
contributions to the integrals in (2.1.4) and (2.1.9) are being calculated)
only elements inside the solid box are held in core, stored in a square b+1
by b+1 matrix. Once triangle k has been assembled, all rows and columns
numbered less than m_{k+1} have been completely calculated, because m_{k+1} is the
lowest numbered node which belongs to any of the remaining triangles. So
all the subdiagonal elements inside the solid box but outside the dashed box

may be zeroed, using Gauss elimination.
Now rows m_k through $m_{k+1} - 1$ will
not be needed again until the
back substitution phase, so they
are written out onto a (disk) file
and the square in-core matrix is
updated to hold the elements shown
in the dashed box.

Then assembly of triangle k+1 begins,
and its contributions to the elements
in the dashed box are calculated
and so the process continues.

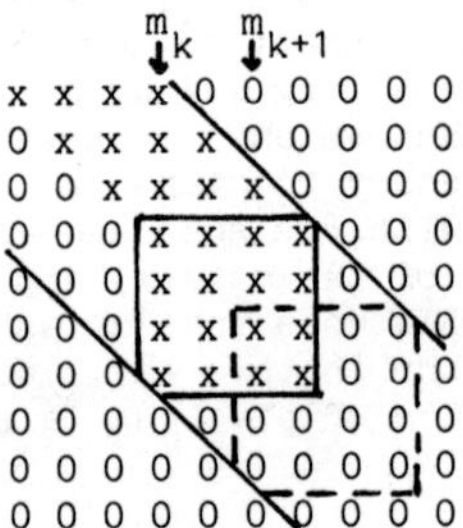

Figure 3.3.1

It may be noticed that when the subdiagonal elements inside the solid but
outside the dashed box are eliminated, some elements in the overlap of the
two boxes are modified by the addition of a multiple of one row to another,
even though there are still triangles remaining which may make a contribution
to them. However, it does not matter if these row multiples are added
before or after they have received their final contribution.

Once the final triangle has been assembled, the rows are read back in, in
reverse order, and back substitution is done.

The PDE/PROTRAN frontal method still takes advantage of symmetry, if present, and in such a case only the diagonal and upper triangle of the b+1 by b+1 box are stored and computed with. Also, PDE/PROTRAN reserves a certain amount of storage as a buffer area, so that rows are not written to and read from the file one at a time, but only when the buffer is full or empty, respectively. This greatly cuts down on the input/output costs at some installations.

Since b is proportional to $N^{0.5}$, the core storage required by the frontal method is quite small, of order $O(N)$, and its efficient organization results in only moderately increased costs compared to the band solver.

In the frontal method, assembly and elimination are intimately intertwined, and cannot be conveniently separated into different subprograms.

3.4 **The Lanczos Solver**

The band and frontal solvers require $O(N^2)$ operations to solve the linear system, and since even within the skyline most of the elements of the matrix are zeros, it might be suspected that full advantage has not yet been taken of sparseness. In fact, there are a number of iterative methods which can solve many of the linear systems of interest here in $O(N^{1.5})$ time, with $O(N)$ storage. One such iterative method, which does take full advantage of sparseness, will be presented here.

Iterative methods, as opposed to direct methods based on Gaussian elimination, generally do not find the exact solution in a finite number of steps, but attempt to decrease the error each step until the approximate solution is sufficiently accurate. Many such methods are known, and the most popular is probably the method of successive over-relaxation, SOR. (See reference 3 and Problem 3.1). Most are ideally designed for positive definite linear systems, and fail on some nonsymmetric problems.

The iterative method available as an option in PDE/PROTRAN is based on the Lanczos, or "bi-conjugate gradient" method {4}, which is described below. The convention of using boldface type for vectors is abandoned in this section.

$$x_0 = \text{initial guess at solution of } Ax=b$$

$$r_0 = b-Ax_0$$

$$\tilde{r}_0 = \text{may be chosen arbitrarily, in principle}$$

$$p_0 = r_0$$

$$\tilde{p}_0 = \tilde{r}_0$$

(3.4.1)

$n=0\ldots N-1$

$$\lambda_n = \tilde{r}_n^T r_n / \tilde{p}_n^T A p_n$$

$$x_{n+1} = x_n + \lambda_n p_n$$

$$r_{n+1} = r_n - \lambda_n A p_n$$

$$\tilde{r}_{n+1} = \tilde{r}_n - \lambda_n A^T \tilde{p}_n$$

$$\alpha_n = -\tilde{p}_n^T A r_{n+1} / \tilde{p}_n^T A p_n$$

$$\tilde{\alpha}_n = -\tilde{r}_{n+1}^T A p_n / \tilde{p}_n^T A p_n$$

$$p_{n+1} = r_{n+1} + \alpha_n p_n$$

$$\tilde{p}_{n+1} = \tilde{r}_{n+1} + \tilde{\alpha}_n \tilde{p}_n$$

If A is symmetric, and $\tilde{r}_0$ is set to r_0, then $\tilde{r}_n=r_n$, $\tilde{p}_n=p_n$ and the algorithm reduces to the usual conjugate gradient method.

If the algorithm does not "breakdown," that is, if $\bar{p}_n{}^T A p_n$ is never zero and $\lambda_n = 0$ only if $r_n = 0$, the following properties can be proven through a massive simultaneous induction process, carried out in Appendix 1:

$$\bar{r}_j{}^T r_i = 0 \qquad \text{when } i \neq j$$

$$(3.4.2) \qquad \bar{p}_j{}^T A p_i = 0 \qquad \text{when } i \neq j$$

$$\bar{p}_j{}^T r_i = 0 \qquad \text{when } i > j$$

Again assuming no breakdown, the sequence $\bar{p}_i, i=0,1\ldots N-1$, is linearly independent, since if:

$$\sum a_i \bar{p}_i = 0$$

$$\text{then} \quad \sum a_i \bar{p}_i{}^T A p_\ell = 0$$

$$a_\ell \bar{p}_\ell{}^T A p_\ell = 0$$

$$a_\ell = 0$$

Also notice that $r_i = b - A x_i$, so that r_i represents the residual, since by induction:

$$r_0 = b - A x_0$$

$$\text{and if} \quad r_n = b - A x_n$$

$$\text{then} \quad r_{n+1} = r_n - \lambda_n A p_n = b - A x_n - \lambda_n A p_n =$$

$$b - A(x_n + \lambda_n p_n) = b - A x_{n+1}$$

So the last formula in (3.4.2) means that the residual r_N is orthogonal to the N linearly independent vectors $\bar{p}_0 \ldots \bar{p}_{N-1}$, and hence it must be zero, that is, x_N must be the exact solution to $Ax=b$, at least if infinite precision is used.

The formula for $\tilde{\alpha}_n$ gives:

$$\tilde{\alpha}_n = -\bar{r}_{n+1}{}^T A p_n / \bar{p}_n{}^T A p_n$$

$$= -\bar{r}_{n+1}{}^T (r_n - r_{n+1}) / \lambda_n / \bar{p}_n{}^T A p_n$$

$$= \bar{r}_{n+1}{}^T r_{n+1} / \bar{r}_n{}^T r_n$$

and a similar expansion for α_n leads to the same result, so that each may be replaced by the above quantity in (3.4.1).

Rather than use (3.4.1) as it stands, however, PDE/PROTRAN "preconditions" the system $Ax=b$ or $b-Ax=0$ by multiplying through by D^{-1}, where D is the "block diagonal" of A. That is, if a system of m PDEs is being solved, and A is thought of as having elements which are m by m matrices, then D is the diagonal of A.

So the linear system being solved is $D^{-1}b-D^{-1}Ax = 0$. Replacing A by $D^{-1}A$ and b by $D^{-1}b$ in (3.4.1), and making the change of variables $\tilde{R}_n = D^{-T}\tilde{r}_n$, $\tilde{P}_n = D^{-T}\tilde{p}_n$, the algorithm becomes:

$$x_0 = \text{initial guess}$$

$$r_0 = D^{-1}b - D^{-1}Ax_0$$

$$\tilde{R}_0 = D^{-T}b - D^{-T}Ax_0$$

$$p_0 = r_0$$

$$\tilde{P}_0 = \tilde{R}_0$$

$$(3.4.3) \qquad \begin{aligned} &\lambda_n = \tilde{R}_n^T Dr_n / \tilde{P}_n^T Ap_n \\[4pt] &x_{n+1} = x_n + \lambda_n p_n \\[4pt] &r_{n+1} = r_n - \lambda_n D^{-1}Ap_n \\[4pt] &\tilde{R}_{n+1} = \tilde{R}_n - \lambda_n D^{-T}A^T\tilde{P}_n \\[4pt] &\alpha_n = \tilde{R}_{n+1}^T Dr_{n+1} / \tilde{R}_n^T Dr_n \\[4pt] &p_{n+1} = r_{n+1} + \alpha_n p_n \\[4pt] &\tilde{P}_{n+1} = \tilde{R}_{n+1} + \alpha_n \tilde{P}_n \end{aligned}$$

$n=0 \ldots N-1$

Now (3.4.3) is the algorithm as actually implemented by PDE/PROTRAN. The stopping criterion is:

$$||r_n||_\infty \leq \max(10^{-8}, \varepsilon^{0.75}) \max(||r_0||_\infty, ||x^k||_\infty)$$

where x^k is the previous Newton iterate (see 3.1.4) and ε is the machine relative precision. Since $x_0 = 0$, $||r_0||_\infty \leq ||D^{-1}A||_\infty ||d^k||_\infty$ where d^k is the solution of the linear system (3.1.4). Then:

$$||A^{-1}b-x_n||_\infty \leq ||A^{-1}D||_\infty ||r_n||_\infty \leq \delta\kappa(D^{-1}A) \max(||d^k||_\infty, ||x^k||_\infty / ||D^{-1}A||_\infty)$$

where $\delta = \max(10^{-8}, \varepsilon^{0.75})$ and $\kappa(D^{-1}A) = ||D^{-1}A||_\infty ||A^{-1}D||_\infty$ is the condition number. Since $||D^{-1}A||_\infty = O(1)$, the iteration is designed to stop when the error is small compared to either the previous Newton iterate or the Newton step size.

Although only the residual is controlled, and the error is amplified by the condition number, notice that this is also the case when direct methods are used.

The iteration also stops, unsuccessfully, if convergence has not occurred by the (N+1)st iteration.

The arbitrary initial vector $\tilde{R}_0$ is chosen in such a way that if A (and thus D) is symmetric, $r_n=\tilde{R}_n$ and $p_n=\tilde{P}_n$ and (3.4.3) becomes:

$$x_0 = \text{initial guess}$$

$$r_0 = D^{-1}b - D^{-1}Ax_0$$

$$p_0 = r_0$$

(3.4.4)

$$\lambda_n = r_n^T D r_n \ / \ p_n^T A p_n$$

$$x_{n+1} = x_n + \lambda_n p_n$$

n=0...N-1

$$r_{n+1} = r_n - \lambda_n D^{-1} A p_n$$

$$\alpha_n = r_{n+1}^T D r_{n+1} \ / \ r_n^T D r_n$$

$$p_{n+1} = r_{n+1} + \alpha_n p_n$$

When A (and thus D) is also positive definite, the change of variables $X_n = D^{0.5}x_n$, $R_n = D^{0.5}r_n$, $P_n = D^{0.5}p_n$ leads to the usual conjugate gradient algorithm (cf 3.4.1) applied to the system $QX = D^{-0.5}b$, where $Q = D^{-0.5}AD^{-0.5}$. If A is positive definite, so is Q, and thus in the important case of a positive definite linear system, the PDE/PROTRAN algorithm cannot break down, since it is well known that the usual conjugate gradient algorithm cannot break down for a positive definite matrix (assuming infinite precision).

In fact, all of the nice properties of the conjugate gradient method apply in this case. In particular (using 3.4.1 when $A=A^T=Q$):

$$R_{n+1}^T Q^{-1} R_{n+1} = (R_n - \lambda_n Q P_n)^T Q^{-1} (R_n - \lambda_n Q P_n)$$

$$= R_n^T Q^{-1} R_n - \lambda_n P_n^T R_n - \lambda_n R_n^T P_n + \lambda_n^2 P_n^T Q P_n$$

$$= R_n^T Q^{-1} R_n - 2\lambda_n P_n^T R_n + \lambda_n R_n^T R_n$$

$$= R_n^T Q^{-1} R_n - \lambda_n R_n^T R_n + 2\lambda_n R_n^T (R_n - P_n)$$

$$= R_n^T Q^{-1} R_n - \lambda_n R_n^T R_n + 2\lambda_n R_n^T (-\alpha_{n-1} P_{n-1})$$

$$= R_n^T Q^{-1} R_n - (R_n^T R_n)^2 / \ P_n^T Q P_n$$

The last step follows from the fact that R_n is orthogonal to P_{n-1} (3.4.2 in the symmetric case). Since Q, and thus Q^{-1}, is positive definite, $R^TQ^{-1}R$ represents a norm (squared) and so the above shows that, in this norm, the residual decreases every iteration. In fact, it has been shown that while the conjugate gradient method is guaranteed to converge in N iterations to within roundoff error, it actually converges, to a given accuracy, in a number of iterations proportional to the square root of the condition number of the matrix Q {5,Section 7.3}. For two dimensional, second order, positive definite PDEs this condition number can be shown to be proportional to N {8, Section 5.5}.

Now the dominant calculation in (3.4.4) is the matrix-vector multiplication Ap_n. Since the average number of neighbors a node has is independent of N, the total number of nonzero elements in A is O(N), so that, assuming sparsity is taken advantage of in the multiplication, the work per iteration is O(N). Thus the total work for the Lanczos method is $O(N^{1.5})$, at least in the positive definite case, which is better than the estimates for the band and frontal solvers. The storage is clearly O(N), since only the nonzeros of A must be stored.

The fact that the convergence rate depends on the condition number of the matrix also explains the purpose of the D^{-1} preconditioning. If the following measure of the condition of a matrix is used (although the spectral condition number is normally used, the convergence rate can be given in terms of χ):

$$\chi(A) = \prod_{i=1}^{N} (\lambda_{ave}(A)/\lambda_i(A)) = \lambda_{ave}^N(A)/\det(A)$$

it can be shown (Problem 3.3) that $1 \le \chi(A)$ for any matrix with positive, real eigenvalues only (e.g. if positive definite) and that equality holds only when all eigenvalues are equal. It can also be shown (Problem 3.4) that if D is the (block) diagonal of A,

$$\chi(D^{-1}A) = \chi(A) / \chi(D)$$

so that the condition number of $D^{-1}A$ is less than that of A, and much less if the diagonal elements vary over a wide magnitude, as for a graded triangulation. The D^{-1} preconditioned conjugate gradient method (3.4.4) is equivalent to what is often called the "Jacobi Conjugate Gradient" method, which was ranked as the best of 13 iterative and sparse direct methods tested on two-dimensional PDE generated linear systems, in an extensive ELLPACK study {6}.

In the general nonsymmetric case, there is no longer any guarantee that (3.4.3) will not breakdown. Still, for random A and x_0, the probability that $\tilde{P}_n^TAp_n$ will exactly hit zero is zero, and in view of the great success of the "big sky, small plane" theory when applied to Gaussian elimination without pivoting (Section 3.2), it might seem that the Lanczos method is as unlikely to breakdown as the band solver is to hit a zero or nearly zero pivot. In addition, the fact that the residual is orthogonal to a larger and larger subspace suggests that it should be getting smaller and smaller long before the "guaranteed" convergence. The reader may notice the similarity between this and the situation with the Galerkin method itself (Section 2.1). There, in the positive definite case, the error could be well characterized, while in the general case all that could be assured was that the residual is orthogonal to a larger and larger subspace.

In finite precision, unfortunately, if A is ill-conditioned the vectors $\bar{p}_i$ may become nearly linearly dependent so that the residual's orthogonality to these vectors is no guarantee that it is small, and even in the positive definite case convergence is not assured. For large problems, if the iteration has not converged by the N-1st iteration, nothing spectacular happens on the Nth, and even if it did, it would be too late to be competitive with other iterative methods. Because of this, the guaranteed convergence property of the Lanczos and conjugate gradient methods is generally dismissed as being of little practical importance.

This author is not so sure, however, and tends to suspect that the guaranteed convergence in the absence of breakdown is somehow related to the fact that for well-conditioned systems (3.4.3) is very robust. This Lanczos method has been run on a number of highly nonsymmetric and indefinite PDEs and results are reported in {7}. These tests show (3.4.3) to be very robust, perhaps more so than other iterative methods. The results also show the very favorable effect that the D^{-1} preconditioning has on robustness.

In any case, a large portion of the matrices in applications are positive definite, or nearly so, and the preconditioned Lanczos method can reasonably be expected to perform well on these.

In the PDE/PROTRAN implementation, A is never actually assembled. Ap is calculated (similarly for $A^T p$) as $\sum A_k p$ where the A_k are the small dense element matrices (the portion of A which comes from integrating over element k), of rank nm, where n is the number of nodes per element (6,10 or 15) and m is the number of PDEs. This approach has two advantages. First, the programming is greatly simplified, since all of the usual sparse matrix formats are very cumbersome to use. Secondly, the multiplication of a vector by an arbitrary sparse matrix does not vectorize on a vector computer and is slowed even on a scalar computer by the need to locate the next non-zero element before each multiplication. (Multiplication by A^T is even worse.) When the unassembled form of A is used, however, the matrix-vector product can be done efficiently on a vector computer, with operations on vectors of length $O(N)$.

Table 3.4.1 and Figure 3.4.1 compare the three PDE/PROTRAN linear system solvers on two typical test problems. The problems are:

(I) $\nabla^2 u + 1 - x^2 - y^2 = 0$ in the unit circle
    $u = 0$ on the boundary

Quadratic elements were used for this problem, which was run on an IBM 4341 in double precision.

(II) $\qquad \nabla^2 u - 2\, \nabla^2 v + 2\, e^{x+y} = 0$ in the unit circle

$$3\, \nabla^2 u - \nabla^2 v - 4\, e^{x+y} = 0$$

$$u = v = e^{x+y} \quad \text{on the boundary}$$

Quartic elements were used, and the tests were run on a Data General MV10000 in double precision.

Table 3.4.1

Problem	I	I	I	I	II	II	II	II
Triangles	500	1000	2000	4000	50	100	200	400
Unknowns	953	1905	3905	7809	754	1506	3106	6210
Lanczos iterations	99	119	167	221	112	147	202	290
storage (thousands of words)								
Lanczos	18	36	72	144	45	90	180	360
Frontal	14	28	55	116	102	174	371	670
Band	205	596	1714	5006	448	1178	3559	9601
CPU time (seconds)								
Lanczos	56	134	381	1022	101	274	748	2096
Frontal	82	266	947	3298	152	*	*	*
Band	31	*	*	*	88	320	1374	*

* = insufficient resources (core or disk storage)

In these tests, and in other more difficult tests, the $O(N^{0.5})$ iteration count, and thus the $O(N^{1.5})$ operation count, is clearly observed, for the Lanczos method. The average slope of the curves log(iterations) vs. log(unknowns) is 0.39 for the first problem and 0.45 for the second.

Nevertheless the gain in speed in not as spectacular as might be supposed by comparing the asymptotic orders. Even with 7809 unknowns, the Lanczos method is only about 3 times as fast as the frontal solver. Thus the band and frontal methods remain viable options, until the size of the problem is very large.

The Lanczos method is competitive with the frontal method in storage efficiency and both are much better than the in-core band solver. In the problem with 7809 unknowns, they require about 40 times less fast memory. However, the frontal method requires a large amount of disk space and input/output operations may be expensive, so the iterative method has the advantage overall in the category of storage requirements.

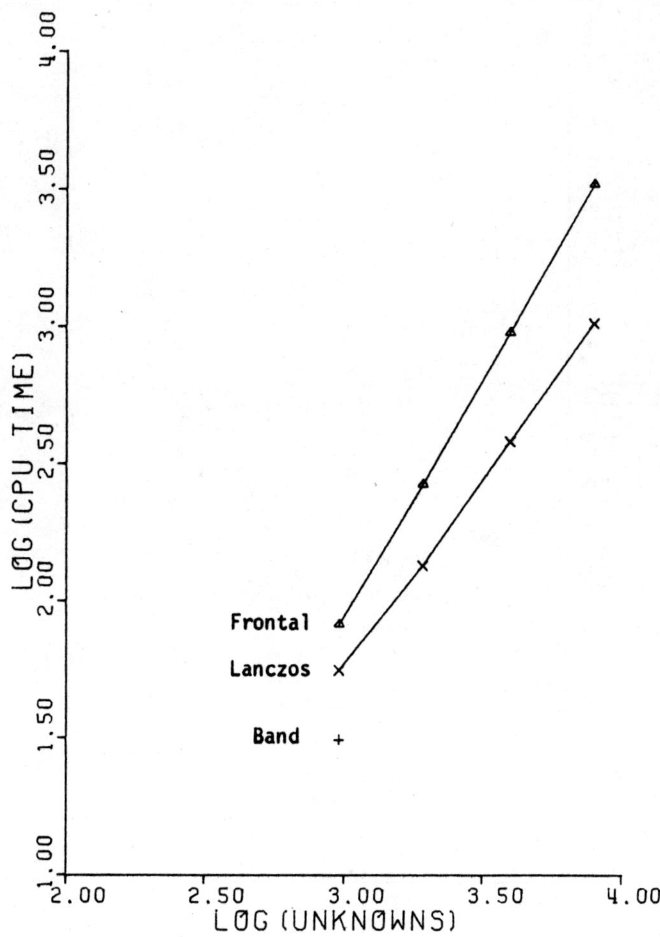

Figure 3.4.1a   CPU Time for Three Linear Equation Solvers (Problem I)

71

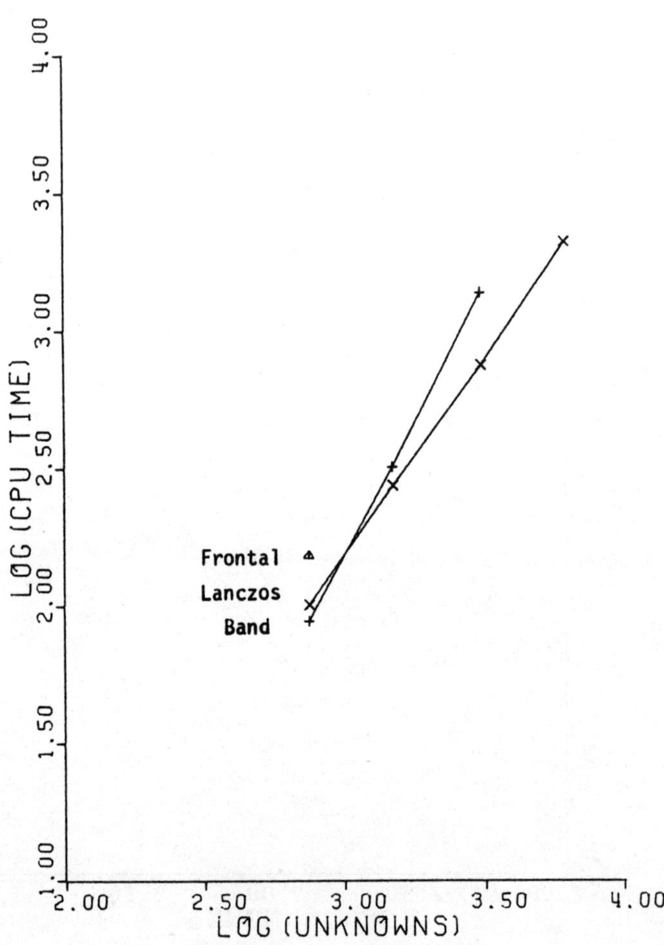

Figure 3.4.1b   CPU Time for Three Linear Equation Solvers (Problem II)

72

3.5 **Exercises** (* = requires use of PDE/PROTRAN)

3.1 The Jacobi iterative method for solving a linear system is defined by solving the ith equation for the ith unknown and iterating with this formula. Show that this is equivalent to replacing the Jacobian in the Newton-Raphson formula (3.1.1) by its diagonal:

$$\mathbf{x}^{k+1} = \mathbf{x}^k - \omega \, D_J(\mathbf{x}^k)^{-1}\mathbf{f}(\mathbf{x}^k)$$

where $\omega=1$. If D_J and $\mathbf{f}$ are evaluated, not using $\mathbf{x}^k$, but using the latest calculated values of x_j (when x_i is calculated, new values of $x_1...x_{i-1}$ are available) the method is called the Gauss-Siedel iteration. If, further, ω is allowed to take values different from 1 the method is called successive overrelaxation (SOR) and if ω is chosen optimally (generally slightly less than 2, for PDE applications) this method is competitive in speed with other iterative methods. Notice that it can be applied to nonlinear systems as well, in which case it is called "nonlinear overrelaxation."

3.2 Consider the unit square, divided into 4LM triangles as shown below:

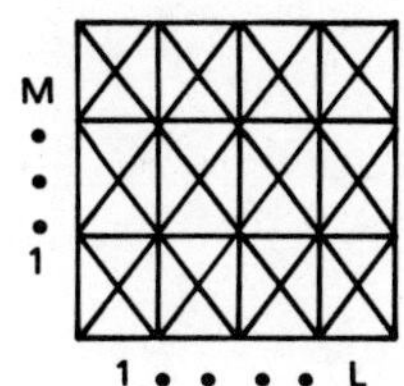

If piecewise linear elements on triangles are used and the Cuthill McKee algorithm is used to number the nodes beginning at the origin:

a. Approximately how many unknowns are there?

b. Approximately what is the half band-width?

c. For two dimensional problems, suppose $M=L\to\infty$. Show that the total work for the band solver is $O(N^2)$.

d. If PDE/PROTRAN is used to solve a one-dimensional problem, where all functions are independent of y, M may be taken to be 1 while $L\to\infty$. Show that the total work for the band solver is $O(N)$ in this case.

e. Would you expect the asymptotic operation count orders for problems c and d to change if higher order elements are used?

3.3 If a condition measure for matrices with all real positive eigenvalues λ_i is defined by:

$$\chi(A) = \prod_{i=1}^{N} (\lambda_{ave}(A)/\lambda_i(A))$$

show that $1 \leq \chi(A)$ and that equality holds only when all eigenvalues are equal. (Hint: If any eigenvalue approaches 0, $\chi(A) \to +\infty$. Thus the minimum value of $\chi(A)$ for all positive λ_i must occur at a point where the gradient of χ, or the gradient of $\log(\chi)$, is zero.)

3.4 Show that if D is the (block) diagonal of A (see Problem 3.3):

$$\chi(D^{-1}A) = \chi(A) / \chi(D)$$

(Hint: The sum of the eigenvalues of a matrix is equal to its trace and the product of its eigenvalues is equal to its determinant.)

*3.5 The equations for a long beam clamped at one end and subjected to a shearing force at the other may be written (See 1.1.3 and 1.1.4):

$$(\sigma_{11})_x + (\sigma_{12})_y = 0$$

$$(\sigma_{12})_x + (\sigma_{22})_y = 0$$

where

$$\sigma_{11} = E(u_x + \nu v_y)/(1-\nu^2)$$

$$\sigma_{22} = E(v_y + \nu u_x)/(1-\nu^2)$$

$$\sigma_{12} = 0.5E(u_y + v_x)/(1+\nu)$$

with boundary conditions:

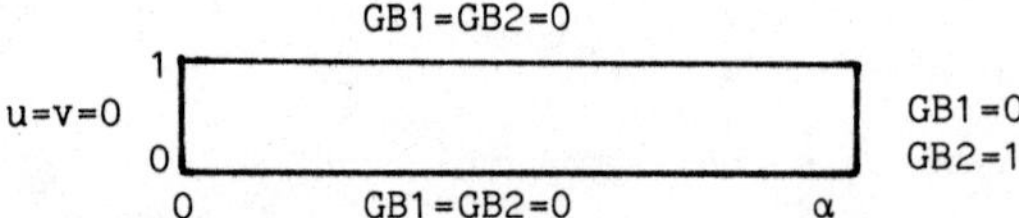

With E=1 and ν=0.2, use PDE/PROTRAN to solve this problem for several values of α, using NTRIANGLES = 4, BAND, and allowing several iterations of Newton's method for this linear problem. For large α, the problem is ill-conditioned, and roundoff error should appear, but be eliminated by the extra iterations.

*3.6 Use the Lanczos (CONJUGATEGRADIENT keyword) option to solve the linear systems generated in Problem 3.5. Note that as α becomes large, the iteration slows and fails to converge eventually, due to the ill-conditioned nature of the problem.

*3.7   Consider the nonlinear minimal surface problem (Example 1, Section 1.1)
       with T=1, f=0, in the first quadrant of the unit circle, with the
       boundary condition:

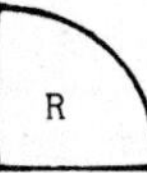

$$u = 0.5Arcsin(1-2sinh^2x\ sinh^2y)$$

       With initial condition u=1.0, try to solve this problem with and
       without steplimitation, using PDE/PROTRAN.

*3.8   Solve the preceding problem using continuation but no steplimitation.
       What is a reasonable parameterization?

*3.9   Solve the highly nonsymmetric problem in Problem 2.11 using the Lanczos
       method to solve the linear system.

*3.10 Solve the fluid flow Problem 2.14 again, this time with ρ=2 (Reynold's
       number approximately 200). Now the problem is nonlinear and it may be
       necessary to parameterize the nonlinear terms or supply very good
       initial values. Plot the velocity field and note that it is no longer
       symmetric about x=0. See Figure 3.5.1.

References

1.   E. Cuthill and J. McKee, "Reducing the Bandwidth of Sparse Symmetric
     Matrices," Proc. ACM Nat. Conf. 157-172 (1969)

2.   B. M. Irons, "A Frontal Solution Program for Finite Element Analyses,"
     Int. J. for Num. Meth. in Eng. 2, 5-32 (1970)

3.   G. E. Forsythe and W. R. Wasow, Finite Difference Methods for Partial
     Differential Equations, John Wiley and Sons, Inc., New York (1960)

4.   K. C. Jea and D. M. Young, "On the Simplification of Generalized
     Conjugate Gradient Methods for Nonsymmetrizable Linear Systems," Linear
     Algebra and Its Applications 52/53: 399-417 (1983)

5.   L. A. Hageman and D. M. Young, Applied Iterative Methods, Academic
     Press, New York (1981)

6.   J. R. Rice, "Performance Analysis of 13 Methods to Solve the Galerkin
     Method Equations," CSD-TR 369, Purdue University, May 1981

7.   E. G. Sewell, "The PDE/PROTRAN Generalized Conjugate Gradient Algorithm,"
     IMSL Tech. Report 8401 (1984)

8.   O. Axelsson and V.A. Barker, Finite Element Solution of Boundary Value
     Problems, Academic Press, New York, 1984.

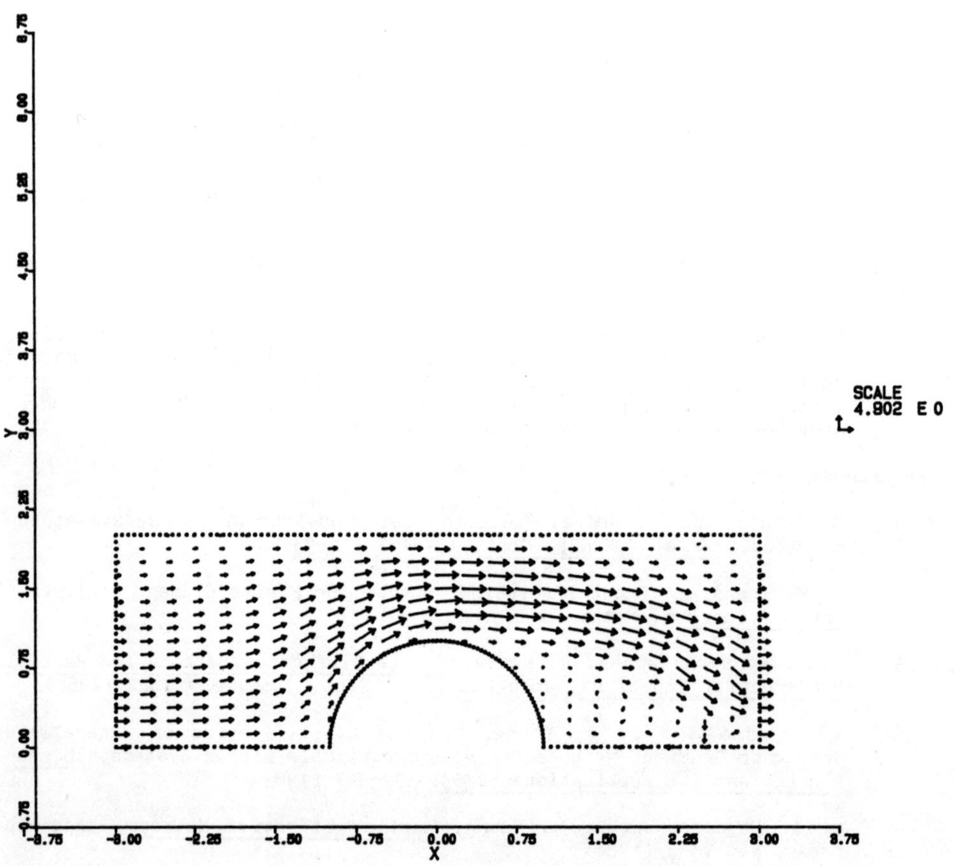

Figure 3.5.1 Velocity Field at Reynold's Number 200 (Problem 3.10)

CHAPTER 4. PARABOLIC PROBLEMS

4.1 The Time Discretization

The form of the time-dependent PDE system solved by PDE/PROTRAN (Section 1.5) is:

$$C(x,y,t,u)u_t = A_x(x,y,t,u,u_x,u_y)+B_y(x,y,t,u,u_x,u_y)+F(x,y,t,u,u_x,u_y) \text{ in } R$$

$$(4.1.1) \qquad u = FB(x,y,t) \qquad \text{on } \partial R_1$$

$$An_x + Bn_y = GB(x,y,t,u) \qquad \text{on } \partial R_2$$

$$u = UO(x,y) \qquad \text{at } t=t_0$$

where R is a general two dimensional region, ∂R_1 and ∂R_2 are disjoint parts of the boundary, and C is a diagonal m by m matrix (m=number of PDEs).

The techniques used to solve (4.1.1) are just extensions of those used to solve (2.1.1) so the reader is advised to study Chapters 2-3 before proceeding.

A solution will be sought of the form:

$$(4.1.2) \qquad u_G(x,y,t) = \phi_0(x,y,t) + \sum_{j=1}^{N} a_j(t)\phi_j(x,y)$$

where the ϕ_j are the same piecewise polynomial functions as in Section 2.2 and

$$\phi_0(x,y,t) = \sum_{i=-1}^{-N1} FB(x_i,y_i,t)\phi_i(x,y)$$

so that u_G satisfies the first boundary condition, at least at the nodes (x_i,y_i) of ∂R_1.

The approximate solution is required to interpolate to UO at all the nodes when $t=t_0$, so that $a_j(t_0) = UO(x_j,y_j)$. Presumably $UO(x,y) = FB(x,y,t_0)$ on ∂R_1; otherwise the boundary and initial conditions are inconsistent.

Now the "weak formulation" of (4.1.1) is obtained by multiplying the PDE and second boundary condition by an arbitrary smooth function ϕ which satisfies $\phi=0$ on ∂R_1 and integrating over R and ∂R_2 respectively. After integration by parts (cf. 2.1.3):

$$\iint_R (-Cu_t\phi + F\phi - A\phi_x - B\phi_y) \, dxdy + \int_{\partial R_2} GB\phi \, ds = 0$$

As in the steady-state case it is not generally possible to find a $\mathbf{u_G}$ of the form (4.1.2) which satisfies the weak formulation for arbitrary ϕ, so it is only required that:

(4.1.3)

$$\iint_R \left\{-C_G\left[(\boldsymbol{\phi_0})_t + \sum_{j=1}^N a_j'(t)\phi_j\right]\phi_k + F_G\phi_k - A_G(\phi_k)_x - B_G(\phi_k)_y\right\} dxdy + \int_{\partial R_2} GB_G\phi_k\, ds = 0\ ,$$

$$k=1\ldots N$$

where the subscript G denotes evaluation at $\mathbf{u_G}$.

Now (4.1.3) is a system of ordinary differential equations (ODEs) for the unknown functions $\mathbf{a_j}(t)$. This ODE system is almost always solved by finite differences, as the primary advantages of finite elements disappear when the "region" is a one dimensional interval $t_0 \leq t \leq t_f$.

PDE/PROTRAN approximates $\mathbf{a_j}'(t)$ (and similarly for $(\boldsymbol{\phi_0})_t$) by the simple difference $(\mathbf{a_j}(t_{n+1}) - \mathbf{a_j}(t_n))/dt_n$, where $t_0, t_1 \ldots t_f$ form an increasing sequence of time values, and $dt_n = t_{n+1} - t_n$. It replaces $\mathbf{u_G}$ in (4.1.3) by $\mathbf{u_G}(x,y,t_n) + \alpha(\mathbf{u_G}(x,y,t_{n+1}) - \mathbf{u_G}(x,y,t_n))$ and t by $t_n + \alpha dt_n$. This gives ($a_j^n = \mathbf{a_j}(t_n)$):

(4.1.4)

$$\iint_R \left\{-C_\alpha\left[\boldsymbol{\phi_0}(x,y,t_{n+1}) - \boldsymbol{\phi_0}(x,y,t_n) + \sum_{j=1}^N (a_j^{n+1} - a_j^n)\phi_j\right]\phi_k/dt_n\right.$$

$$\left. + F_\alpha\phi_k - A_\alpha(\phi_k)_x - B_\alpha(\phi_k)_y\right\} dxdy + \int_{\partial R_2} GB_\alpha\phi_k\, ds = 0$$

where the subscript α denotes evaluation at:

$$t = t_n + \alpha dt_n$$

$$u = \boldsymbol{\phi_0}(x,y,t_n) + \alpha(\boldsymbol{\phi_0}(x,y,t_{n+1}) - \boldsymbol{\phi_0}(x,y,t_n))$$

$$+ \sum_{j=1}^N (a_j^n + \alpha(a_j^{n+1} - a_j^n))\, \phi_j(x,y)$$

and similarly for $\mathbf{u_x}, \mathbf{u_y}$. α is normally chosen to be 1 (backward difference method) or 0.5 (Crank-Nicolson method). Clearly $\alpha=0.5$ will lead to an $O(dt^2)$ time discretization (cf. Problem 4.2), since everything is centered at $t_n + 0.5dt_n$, whereas the backward difference method will only give $O(dt)$.

Assuming $\mathbf{u_G}(x,y,t_n)$ has been calculated and the solution at t_{n+1} is now desired, (4.1.4) constitutes a system of Nm algebraic equations for the Nm unknowns, a_j^{n+1}, $j=1\ldots N$. A form of Newton's method is again used to solve this system.

The Jacobian "element" G_{kj} (really an m by m matrix) can be calculated by taking the partial derivative of (4.1.4) with respect to a_j^{n+1}. After considerable effort, it is found that:

$$(4.1.5) \quad G_{kj} = \iint_R \{-[\boldsymbol{\phi_0}(x,y,t_{n+1})-\boldsymbol{\phi_0}(x,y,t_n)+ \sum (a_j^{n+1}-a_j^n)\phi_j] \ (\partial C/\partial u)_\alpha \alpha \phi_j \phi_k/dt_n$$

$$- C_\alpha \ \phi_j \phi_k/dt_n\} \ dxdy + \alpha(J_{kj})_\alpha$$

where the J_{kj} are given by (2.1.9), evaluated at the same values of t,u,u_x,u_y as in (4.1.4). $\partial C/\partial u$ represents the Jacobian of C, with C thought of as a vector rather than a diagonal matrix, and the first term in brackets is considered to be a diagonal matrix.

Now the values of a_j^n from the previous time step are used as the starting values for Newton's iteration. Thus on the first Newton iteration the Jacobian is evaluated at $a_j^{n+1}=a_j^n$ so that:

$$G_{kj} = \iint_R \{-[\boldsymbol{\phi_0}(x,y,t_{n+1})-\boldsymbol{\phi_0}(x,y,t_n)](\partial C/\partial u)_n \alpha \phi_j \phi_k/dt_n - C_n \ \phi_j \phi_k/dt_n\} \ dxdy + \alpha(J_{kj})_n$$

where the subscript n indicates that the functions are evaluated at $t = t_n + \alpha dt_n$ and:

$$\mathbf{u} = \boldsymbol{\phi_0}(x,y,t_n)+\alpha(\boldsymbol{\phi_0}(x,y,t_{n+1})-\boldsymbol{\phi_0}(x,y,t_n)) + \sum_{j=1}^N a_j^n \phi_j(x,y)$$

Now C is rarely a function of **u** in applications (usually C=I) and so usually the first term in G_{kj} vanishes. In any case this term is zero at every node in R, because $\boldsymbol{\phi_0}$ is zero at all nodes except those on ∂R_1 while ϕ_j and ϕ_k are both zero at all nodes on ∂R_1. Thus if the integration points were taken to be just the nodes, the integral of this first term would be zero. So PDE/PROTRAN simply ignores this term, considering that in the worst case there will be some loss of accuracy in the Jacobian (not in the function vector) due to the fact that the numerical integration scheme which evaluates the integrand at the nodes is not as accurate as the scheme used otherwise (for quadratics it **is** as accurate as the normal three point scheme). And as will be seen later, a small loss of accuracy in the Jacobian only slows the Newton convergence and does not affect the accuracy in the solution.

Thus the Jacobian used by PDE/PROTRAN, for the first Newton iteration, has elements:

$$(4.1.6) \quad G_{kj} = \iint_R -C_n \phi_j \phi_k/dt_n \ dxdy + \alpha(J_{kj})_n$$

If the PDE system is linear, of course, only one iteration is necessary. Now for reasons which will be discussed in Section 4.3, it is not practical to update the Jacobian each iteration. The Jacobian (4.1.6), appropriate for the first iteration, or even the Jacobian appropriate for an earlier time step, will be used on subsequent iterations. Thus it is only a pseudo-Newton fixed point iteration that is actually used, and even linear problems require more than one iteration. This will be discussed further in Section 4.3.

Now it can be shown that if the time step is constant, $dt_n = dt$, the time discretization error for the Crank-Nicolson method, that is, the difference between $\mathbf{u_G}(x,y,t_n)$ and the exact solution of the ODE system (4.1.3), $\mathbf{u}^*(x,y,t_n)$, is given by (cf. Problem 4.2):

$$\mathbf{u_G}(x,y,t_n) - \mathbf{u}^*(x,y,t_n) = \mathbf{P}(x,y,t_n)\, dt^2 + O(dt^4)$$

and for the backward difference method, by:

$$\mathbf{u_G}(x,y,t_n) - \mathbf{u}^*(x,y,t_n) = \mathbf{Q}(x,y,t_n)\, dt + O(dt^2)$$

where $\mathbf{P}$ and $\mathbf{Q}$ are, of course, not known explicitly.

Thus a Richardson extrapolation can be done to double the order of convergence of the time discretization error. If the problem is solved twice, once with stepsize dt, and the next with dt/2, with no other changes between runs, then for Crank-Nicolson:

$$\mathbf{u_G}(x,y,t_n;dt) - \mathbf{u}^*(x,y,t_n) = \mathbf{P}(x,y,t_n)\, dt^2 + O(dt^4)$$

$$\mathbf{u_G}(x,y,t_n;dt/2) - \mathbf{u}^*(x,y,t_n) = \mathbf{P}(x,y,t_n)\, dt^2/4 + O(dt^4)$$

so

$$\{4/3\ \mathbf{u_G}(x,y,t_n;dt/2) - 1/3\ \mathbf{u_G}(x,y,t_n;dt)\} - \mathbf{u}^*(x,y,t_n) = O(dt^4)$$

and so the term in brackets is an $O(dt^4)$ accurate estimate of the solution.

Similarly, for the backwards difference method:

$$\{2\ \mathbf{u_G}(x,y,t_n;dt/2) - \mathbf{u_G}(x,y,t_n;dt)\} - \mathbf{u}^*(x,y,t_n) = O(dt^2)$$

A very important by-product of the Richardson extrapolation is that it produces a reliable bound on the global time-discretization error, something almost impossible to obtain without actually solving the problem twice, as done here. For Crank-Nicolson, for example:

$$\{1/3\ \mathbf{u_G}(x,y,t_n;dt) - 1/3\ \mathbf{u_G}(x,y,t_n;dt/2)\} = \mathbf{P}(x,y,t_n)\, dt^2/4 + O(dt^4)$$

and the right hand side is almost equal to the error in $\mathbf{u_G}(x,y,t_n;dt/2)$. Presumably the error after extrapolation is substantially smaller.

The above assumes a constant time step size, but PDE/PROTRAN allows a variable step size. Like the grading of the space mesh, the grading of the time spacing is not done adaptively, but is specified a priori by the user. This a priori specification, however, is what makes possible the Richardson extrapolation, even when a variable time step is used. A function DTINV(t) (analogous to the D3EST(x,y) function), is supplied by the user along with an average time step size, dt. The time grid points, t_n, are defined by requiring that the change of time variable:

$$s(t) = t_0 + (t_f - t_0) \int_{t_0}^{t} DTINV(r)\, dr \; / \; \int_{t_0}^{t_f} DTINV(r)\, dr$$

result in a constant step size in s, i.e., $s(t_n) = t_0 + n\, dt$. Note that the step size dt_n is inversely proportional to DTINV(t), so DTINV(t) should be large where dt_n is to be small.

It can easily be shown that if the above change of variables is applied to (4.1.3) and the Crank-Nicolson or backward difference method is used with a constant step size (in s), and then the variable s is transformed back into t, there result equations (4.1.4) again, with one difference. Instead of evaluating the time dependent functions at $t=t_n+\alpha dt_n$, they should be evaluated at $t=t(s_n+\alpha dt) = t(t_0+(n+\alpha)dt)$. This is what is actually used, so that the values of the solution at the t_n will correspond to the values of the solution to the transformed problem at s_n, and the Richardson extrapolation will work correctly. Notice that finding t given s is not as easy as the reverse process.

To check the theoretical time discretization error estimates, consider the problem:

$$u_t = \nabla^2 u - u - 4\, e^{-t} \qquad \text{in the unit square}$$

with boundary conditions:

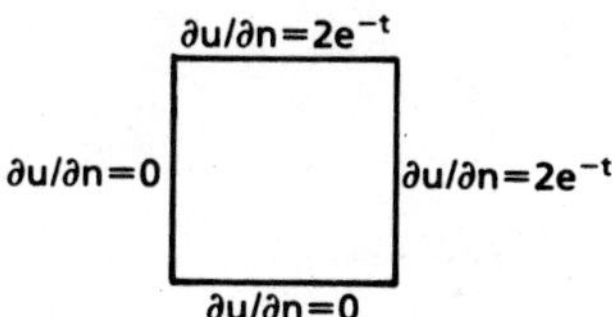

and initial condition:

$$u = 1+x^2+y^2 \qquad \text{when } t=0$$

The exact solution is $u=(1+x^2+y^2)e^{-t}$, chosen because the space discretization error will be zero ($1+x^2+y^2$ can be fit exactly with quadratics) so that all the measured error will be due to the time discretization. The errors at a fixed point $(x,y,t) = (0.5,0.5,1.0)$ with various average step sizes DT (a variable step size is used) and both finite difference options, with and without extrapolation, were calculated and recorded in Table 4.1.2. Table 4.1.1 shows the PDE/PROTRAN input to solve one of these problems. To do an extrapolation, the input in Table 4.1.1 is repeated with only two changes: DT, the average time step size, is doubled and SAVEFILE=FSTRUN is replaced by EXTRAPOLATE=FSTRUN.

Table 4.1.1

```
$       PDE2D
C   PDE AND AUXILIARY CONDITIONS
        C = 1.0
        A = UX
        B = UY
        F = -U-4.*DEXP(-T)
        XGRID = 0,1
        YGRID = 0,1
        ARCS = (1,2) (3,4)
        GB = (2,2.*DEXP(-T))  (4,2.*DEXP(-T))
        UO = 1.0+X*X+Y*Y
C   OUTPUT ERROR
        UPRINT = U-(1.+X*X+Y*Y)*DEXP(-T)
        NTRIANGLES = 4
C   SAVE SOLUTION FOR EXTRAPOLATION
        SAVEFILE = FSTRUN
        PRINTSOLUTION
        DT = 0.05
        TF = 1.0
        GRIDPOINTS = (3,3)
        OUTFREQUENCY = 5
        CRANKNICOLSON
C   DEFINITION OF DTINV(T)
        DTDENSITY = 1.0+T*T
        PRECISION = DOUBLE
$       END
```

Table 4.1.2

Errors at (0.5,0.5,1.0)

Crank-Nicolson			Backward Difference		
DT	normal	extrapolated	DT	normal	extrapolated
0.1	-.20768E-3	-.87116E-6	0.1	.30599E-1	.69532E-3
0.05	-.52573E-4	-.51531E-7	0.05	.15647E-1	.18034E-3
0.025	-.13182E-4	-.31609E-8	0.025	.79137E-2	.46064E-4
0.0125	-.32978E-5	-	0.0125	.39799E-2	-
ave. exponent of DT	1.99	4.05		0.98	1.96

Notice that in Table 4.1.2, the exponent of DT (average slope of log(error) versus log(DT)) is very close to the theoretical value in all four cases, even though a variable time step is used.

4.2 **Stability**

Since it is well-known that some finite difference methods for initial value problems are unstable, or only conditionally stable, the stability of (4.1.4) is investigated, in the linear case. If $a^n = (a_1^n \ldots a_N^n)$ then the left hand side of (4.1.4) is linear in both a^{n+1} and a^n because the PDE problem is assumed linear. The Jacobian of (4.1.4) with respect to a^{n+1}, the unknown vector, has already been calculated and is given in (4.1.5), but with $\partial C/\partial u = 0$ because of the assumed linearity. That is,

$$G_{kj} = \iint_R -C_\alpha \, \phi_j\phi_k/dt_n \, dxdy + \alpha(J_{kj})_\alpha$$

where the subscript α means that t is evaluated at $t_n + \alpha dt_n$. (G_{kj} is now independent of $\mathbf{u}$, because of the linearity.) The Jacobian of (4.1.4) with respect to a^n is similarly calculated and has components:

$$G'_{kj} = \iint_R C_\alpha \, \phi_j\phi_k/dt_n \, dxdy + (1-\alpha)(J_{kj})_\alpha$$

Since (4.1.4) is really a system of linear equations now, with the above Jacobians with respect to a^{n+1} and a^n, these linear equations can be written:

(4.2.1) $G\, a^{n+1} + G'a^n + b^n = 0$

where G and G' have components as given above and b^n is a vector independent of a^{n+1} and a^n. If a matrix S_C is defined to have

(4.2.2) $(S_C)_{kj} = \iint_R C_\alpha \, \phi_j\phi_k \, dxdy$

and J to have components $(J_{kj})_\alpha$, then (4.2.1) can be written:

$$(-S_C/dt_n + \alpha J)a^{n+1} + (S_C/dt_n + (1-\alpha)J)a^n + b^n = 0$$

or

$$a^{n+1} = (S_C/dt_n - \alpha J)^{-1}(S_C/dt_n + (1-\alpha)J)a^n + (S_C/dt_n - \alpha J)^{-1}b^n$$

$$= B_n a^n + c^n$$

It is known {cf. 1, Section 29.5 and 2, Section 2.7} that a finite difference method expressed as $a^{n+1} = B_n a^n + c^n$ is stable if the absolute values of all eigenvalues μ_k of B_n are less than $1 + Mdt_n$, where M is independent of the spatial grid.

The importance of stability is that if the finite difference method is stable, and is consistent (has a truncation error which goes to zero as $dt \to 0$), convergence to the solution of the ODE system (4.1.3) is assured. For an unstable or conditionally stable method, as the triangulation is refined and as the time stepsize is decreased, the solution may become unbounded, if they are not refined with the "correct" relative speeds.

To see this, recall that the amount by which the exact solution $\mathbf{a}^*(t_n)$ fails to satisfy the difference equations (4.2.1) is called the truncation error, $\mathbf{T^n}$, and for the Crank-Nicolson and backward difference methods it is easily shown to be $O(dt^2)$ and $O(dt)$ respectively (cf. Problem 4.2). Thus if $\mathbf{e^n} = \mathbf{a^n} - \mathbf{a}^*(t_n)$, the error satisfies:

$$G\,\mathbf{e^{n+1}} + G'\mathbf{e^n} + \mathbf{T^n} = \mathbf{0}$$

$$\mathbf{e^{n+1}} = B_n\mathbf{e^n} + (S_C/dt_n - \alpha J)^{-1}\mathbf{T^n}$$

with $\mathbf{e_0} = \mathbf{0}$. Although not necessary, the convergence proof is much simpler if $B_n = B$ is assumed to be constant (and S_C, J, $dt_n = dt$ also), and similiar to a diagonal matrix, $B = P^{-1}DP$. Then the above equation can be solved explicitly for $\mathbf{e^{n+1}}$:

$$\mathbf{e^{n+1}} = \sum_{k=0}^{n} B^k(S_C/dt - \alpha J)^{-1}\mathbf{T^{n-k}}$$

If T represents the maximum of $||\mathbf{T^k}||_\infty$ in the region of interest:

$$||\mathbf{e^{n+1}}||_\infty \le dt\ T\ ||(S_C - \alpha dt J)^{-1}||_\infty \sum_{k=0}^{n} ||B^k||_\infty$$

$$= dt\ T\ ||(S_C - \alpha dt J)^{-1}||_\infty \sum_{k=0}^{n} ||P^{-1}D^kP||_\infty$$

$$\le dtT||(S_C - \alpha dt J)^{-1}||_\infty||P^{-1}||_\infty||P||_\infty \sum_{k=0}^{n} |\mu_{max}(B)|^k$$

$$\le dtT||(S_C - \alpha dt J)^{-1}||_\infty||P^{-1}||_\infty||P||_\infty\ (n+1)(1+Mdt)^{n+1}$$

$$\le(t_{n+1}-t_0)||(S_C - \alpha dt J)^{-1}||_\infty||P^{-1}||_\infty||P||_\infty\exp(M(t_{n+1}-t_0))T \le K(t_{n+1})T$$

Since $K(t_{n+1})$ is bounded and $T \to 0$ as $dt \to 0$, the error goes to zero. This illustrates a fundamental principle of finite differences: stability plus consistency equals convergence.

The eigenvalues of B_n are related to the eigenvalues of the generalized eigenvalue problem $J\mathbf{w} = \lambda S_C\mathbf{w}$, and both have the same eigenvectors. For, suppose $J\mathbf{w_k} = \lambda_k S_C\mathbf{w_k}$. Then:

$$B_n\mathbf{w_k} = \mu_k\mathbf{w_k}$$

$$(S_C/dt_n + (1-\alpha)J)\mathbf{w_k} = \mu_k(S_C/dt_n - \alpha J)\mathbf{w_k}$$

$$(1/dt_n + (1-\alpha)\lambda_k)S_C\mathbf{w_k} = \mu_k(1/dt_n - \alpha\lambda_k)S_C\mathbf{w_k}$$

Assuming S_C is nonsingular (it is positive definite if the diagonal entries of C are positive), and since $\mathbf{w_k} \ne \mathbf{0}$, the eigenvalues of B_n are given by:

(4.2.3)
$$\mu_k = \frac{1+(1-\alpha)\lambda_k dt_n}{1-\alpha\lambda_k dt_n}$$

And so

$$|\mu_k|^2 = \frac{1 + (1-\alpha)^2|\lambda_k|^2 dt_n^2 + 2(1-\alpha)Re(\lambda_k)dt_n}{1 + \alpha^2|\lambda_k|^2 dt_n^2 - 2\alpha Re(\lambda_k)dt_n}$$

Suppose that $0.5 \leq \alpha \leq 1$, which includes both methods of interest here, and that there exists some positive constant M, **independent of the triangulation**, such that $Re(\lambda_k) \leq M$. Suppose further that dt_n is less than the threshold value $1/(4\alpha M)$. Then the denominator above is greater than 0.5, and $1-2\alpha \leq 0$ and therefore:

$$|\mu_k|^2 = 1 + \frac{(1-2\alpha)|\lambda_k|^2 dt_n^2 + 2 Re(\lambda_k)dt_n}{1 + \alpha^2|\lambda_k|^2 dt_n^2 - 2\alpha Re(\lambda_k)dt_n} \leq 1 + 4Mdt_n$$

In summary, then, both the Crank-Nicolson and backward difference methods are stable for any problem for which the real parts of the eigenvalues of $Jw = \lambda S_C w$ have an upper bound as the triangulation is refined. This obviously includes the commonly occurring problems where S_C is positive definite and J is negative definite (Problem 4.1), in which case M may be taken arbitrarily close to zero, so that there is no threshold limit for dt.

But it also includes any linear problem of the form (4.1.1), with smooth coefficients, such that the elements of C are positive and such that the 2m by 2m matrix:

(4.2.4)
$$\begin{bmatrix} A.UX & A.UY \\ B.UX & B.UY \end{bmatrix} + \begin{bmatrix} A.UX & A.UY \\ B.UX & B.UY \end{bmatrix}^T$$

is positive definite (it is always symmetric). Although unnecessary, to simplify the proof it will be assumed that **GB** is not a function of **u**.

Now if $Jw = \lambda S_C w$ it will be shown that $Re(\lambda)$ has an upper bound.

If w^* represents the complex conjugate of **w**, then:

$$2Re(\lambda)w^{*T}S_C w = (\lambda+\lambda^*)w^{*T}S_C w = w^{*T}Jw + w^T Jw^* = w^{*T}(J+J^T)w$$

From (2.1.11), the right hand side can be written (recall that $\mathbf{W} = \sum w_k \phi_k$):

$$\iint_R (\mathbf{w}^*, \mathbf{w}_x^*, \mathbf{w}_y^*)^T \begin{bmatrix} F.U+F.U^T & F.UX-A.U^T & F.UY-B.U^T \\ -A.U+F.UX^T & -A.UX-A.UX^T & -A.UY-B.UX^T \\ -B.U+F.UY^T & -B.UX-A.UY^T & -B.UY-B.UY^T \end{bmatrix} \begin{bmatrix} \mathbf{W} \\ \mathbf{W}_x \\ \mathbf{W}_y \end{bmatrix} dxdy$$

$$\leq \iint_R \{-\sigma(||\mathbf{W}_x||^2 + ||\mathbf{W}_y||^2) + 2||F.UX-A.U^T||\ ||\mathbf{W}||\ ||\mathbf{W}_x||$$

$$+ 2||F.UY-B.U^T||\ ||\mathbf{W}||\ ||\mathbf{W}_y|| + ||F.U+F.U^T||\ ||\mathbf{W}||^2\} dxdy$$

where σ is the smallest eigenvalue of the positive definite matrix (4.2.4), and the norm used is the L_2 norm, for m-vectors and m by m matrices.

Now also, using the definition (4.2.2) for S_C:

$$\mathbf{w}^{*T}S_C\mathbf{w} = \iint_R \mathbf{w}^{*T}C\mathbf{w}\ dxdy \geq \iint_R \sigma_C\ ||\mathbf{w}||^2\ dxdy$$

where σ_C is the smallest eigenvalue of the positive definite diagonal matrix C. Now if the following definitions are made:

$$\alpha_1 = \min_R \sigma_C \qquad \alpha_2 = \max(\max_R ||F.UX-A.U^T||, \max_R ||F.UY-B.U^T||)$$

$$\alpha_3 = \max_R ||F.U+F.U^T|| \qquad \alpha_4 = \min_R \sigma$$

then:

$$2Re(\lambda)\alpha_1 \iint ||\mathbf{w}||^2\ dxdy$$

$$\leq -\alpha_4 \iint ||\mathbf{w}_x||^2+||\mathbf{w}_y||^2\ dxdy + 2\alpha_2 \iint ||\mathbf{w}||(||\mathbf{w}_x||+||\mathbf{w}_y||)dxdy$$

$$+\alpha_3 \iint ||\mathbf{w}||^2\ dxdy$$

$$\leq -\alpha_4 \iint ||\mathbf{w}_x||^2+||\mathbf{w}_y||^2\ dxdy + \alpha_3 \iint ||\mathbf{w}||^2\ dxdy$$

$$+ 2\alpha_2 \{\iint ||\mathbf{w}||^2 dxdy\}^{0.5} \{\iint (||\mathbf{w}_x||+||\mathbf{w}_y||)^2 dxdy\}^{0.5}$$

If z is defined by:

$$z^2 = (\iint ||\mathbf{w}_x||^2 + ||\mathbf{w}_y||^2\ dxdy) / (\iint ||\mathbf{w}||^2\ dxdy)$$

Then:

$$2Re(\lambda) \leq -\alpha_4/\alpha_1\ z^2 + 2^{1.5}\ \alpha_2/\alpha_1\ z + \alpha_3/\alpha_1$$

And so:

$$(4.2.5) \qquad Re(\lambda) \leq -\alpha_4/(2\alpha_1)(z-2^{0.5}\alpha_2/\alpha_4)^2 + (\alpha_3\alpha_4+2\alpha_2^2)/(2\alpha_1\alpha_4)$$

Thus $Re(\lambda)$ clearly has a nonnegative upper bound, since all the α_i are nonnegative and α_1 and α_4 are positive.

Since the Crank-Nicolson method is of higher order accuracy, and always stable, the reader might wonder why the other option is even available. The answer is seen in (4.2.3). If the initial conditions have some high frequency "noise", that is, an appreciable component along $\mathbf{w}_k$ where $|\lambda_k|$ is large, then with $\alpha=1$, μ_k will be small, so that this noise is damped out rapidly. With $\alpha=0.5$, however, it can be seen from (4.2.3) that when $|\lambda_k dt_n|$ is large, μ_k is almost negative one, so the noise is damped out very slowly, and oscillates (Problem 4.4). Thus while both methods are stable, the backward difference method is "more stable."

4.3 **Solving the Linear Equations**

If a true Newton iteration were done to solve the discrete equations (4.1.4) this would mean solving several linear systems, with nonzero structure identical to those solved in the steady state case, each time step. In addition, the time to assemble the Jacobian matrix this many times is by no means negligible in moderate size problems. This would certainly be prohibitively expensive, and would make PDE/PROTRAN uncompetitive on time-dependent problems. Even solving one such system per time step may be far too expensive. Fortunately, for small time steps sizes, the Jacobian does not change much from one step to the next and so an old Jacobian can be used a while, until it gets so out of date that convergence is slow.

For many applications, in fact, the Jacobian does not change at all. If a problem is linear and all functions are independent of time with the possible exception of nonhomogeneous terms, then all of the partial derivatives in (2.1.10) plus those in GB.U will be time-independent, so that by (2.1.9), J will be constant. If, further, C is a function of x and y only, and a constant time step is used, G in (4.1.6) will be constant, and the Jacobian calculated the first step can be used throughout the calculations.

This situation is much more common in applications than might be imagined. PDE/PROTRAN allows the user to identify such cases using the "NOUPDATE" keyword which assures that the Jacobian will be calculated only once, and never updated.

Otherwise, keyword "AUTOUPDATE" requests that the program update only when the "pseudo-Newton" iteration:

$$(4.3.1) \qquad x^{k+1} = x^k - G_0^{-1} f(x^k)$$

converges slowly, where G_0 is the Jacobian evaluated, not at x^k, but at an earlier time.

Convergence of (4.3.1) will certainly no longer be quadratic, as for a true Newton iteration. But if x^* is the true solution:

$$0 = f(x^*) = f(x^k) + G(\xi)(x^*-x^k)$$

where $G(\xi)$ is a matrix whose ith row is the gradient of f_i evaluated at some point ξ_i between x^k and x^*. So,

$$x^{k+1} - x^* = x^k - x^* - G_0^{-1}G(\xi)(x^k-x^*)$$

$$x^{k+1} - x^* = (I-G_0^{-1}G(\xi))(x^k-x^*)$$

Now if dt_n is sufficiently small, very good initial values for the iteration (4.3.1) are available, namely the solution at t_n. Thus x^0 and also x^k will be close to x^*, and so $G(\xi)$ will approximate $G(x^*)$. If, in addition, the Jacobian has not changed too much since G_0 was last updated, then G_0 also approximates $G(x^*)$, so $||I-G_0^{-1}G(\xi)||$ is less than one and the iteration will converge.

Using (4.1.6) and (4.2.2) the Jacobian is written $G_n = -(S_C)_n/dt_n + \alpha J_n$. Since for small dt_n the first term dominates, and since further C is often a function of x and y only (in fact, usually C=I), so that the matrix S_C is constant, when a variable time step is used there is an obvious way to write the pseudo-Newton iteration so as to minimize the frequency of updating necessary. For, if Newton's method is done with updating once each time step,

$$x^{k+1} = x^k - [-(S_C)_n/dt_n + \alpha J_n]^{-1} f(x^k)$$

or

$$x^{k+1} = x^k - dt_n/dt_0 [-(S_C)_n/dt_0 + \alpha dt_n/dt_0 J_n]^{-1} f(x^k)$$

and the matrix in brackets generally varies more slowly than G itself. (Here dt_0 represents the value of dt_n when G was last updated.) Thus the pseudo-Newton iteration actually used replaces the matrix in brackets by G_0:

$$x^{k+1} = x^k - dt_n/dt_0 \; G_0^{-1} f(x^k)$$

or

(4.3.2)
$$G_0(x^{k+1} - x^k) = -dt_n/dt_0 \; f(x^k) = b^k$$

On those pseudo-Newton iterations where an old Jacobian is being used, not only is the work to recalculate the Jacobian avoided, but more importantly, for direct methods the work to solve (4.3.2) is cut dramatically. To see how this is done, consider the band matrix in Figure 4.3.1.

The first time (4.3.2) is solved after a new update, Gaussian elimination is done as described in Section 3.2, except that when row i is multiplied by $-g_{ji}g_{ii}^{-1}$ and added to row j, to eliminate g_{ji}, and the same multiple of b_i is added to b_j, the value of g_{ji} is not actually changed to zero, although it is understood that this position is now zero. The last value of g_{ji} before elimination is allowed to remain there, for use the next time a linear system with the same matrix has to be solved. Notice that neither g_{ii} nor g_{ji} are touched again after this, either in the rest of the elimination or in the back substitution phase.

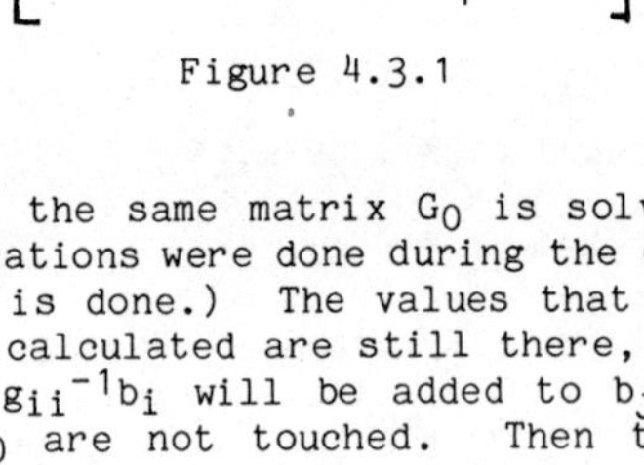

$$
\begin{bmatrix}
g_{11} & & & & & b_1 \\
& \cdot & & & & \cdot \\
& & \cdot & & & \cdot \\
& & g_{ii} & & & b_i \\
& & \cdot & \cdot & & \cdot \\
& & \cdot & & \cdot & \cdot \\
& & g_{ji} & & \cdot & b_j \\
& & & \cdot & & \cdot \\
& & & & \cdot & \cdot
\end{bmatrix}
$$

Figure 4.3.1

Now the next time a linear system with the same matrix G_0 is solved, the matrix contains a record of what row operations were done during the original elimination. (Recall that no pivoting is done.) The values that g_{ii} and g_{ji} had when the multiple $g_{ji}g_{ii}^{-1}$ was calculated are still there, so this multiple can be formed again and $-g_{ji}g_{ii}^{-1}b_i$ will be added to b_j again. However, this time the elements of G_0 are not touched. Then the back substitution is carried out as before, as the superdiagonal elements have not changed since back substitution was last done.

In summary, the first time (4.3.2) is solved, Gaussian elimination is carried out normally, except that the eliminated elements are not actually set to zero. The second and subsequent times, the elimination is carried out normally, except that none of the elements of G_0 are modified--the indicated row operations are done only to **b**. This process is of course almost identical to forming the "LU-decomposition" of G_0.

Symmetry can still be taken advantage of if present, since as shown in Section 3.2, g_{ji} remains equal to g_{ij}^T throughout the elimination, until it is set equal to zero (eliminated). Since, as outlined above, the eliminated elements are not actually zeroed, the matrix remains symmetric throughout the entire process, and so only the upper triangular part needs to be stored and computed with.

If the Jacobian has already been decomposed by Gaussian elimination earlier, this time the forward elimination is applied only to the right hand side vector, and so only Nb operations are required, the same as for the back substitution, where N is the number of unknowns and b is the half band width. So 2Nb operations are required to solve (4.3.2), as opposed to Nb^2 ($0.5Nb^2$ if symmetric) the first time. Since b is proportional to $N^{0.5}$, $2Nb = O(N^{1.5})$.

Since the frontal method is just an out-of-core version of the band solver, all the preceding is also implemented in the frontal solver.

For the Lanczos iterative solver, on the other hand, the only advantage in not updating the matrix every step is the savings in assembling the Jacobian. (Thus the "AUTOUPDATE" keyword is generally not recommended when the Lanczos method is used.)   Iterative methods cannot take such advantage of the fact that the matrix has not changed much. However, this iterative method (and most others) is able to solve the linear system (4.3.2) much more rapidly than the steady state systems of the same size, for two reasons. The first is simply that, for small dt_n, much better initial values are available, so that convergence occurs sooner. This is not as big an advantage as it might seem. If, for example, 10^{-8} relative accuracy is desired and the initial guess is moved 10 times closer to the solution, since convergence is linear the number of iterations will decrease by only about 12.5%.

More important than this, however, is the fact that the Lanczos iteration matrix $D_G^{-1}G$ for time dependent problems is generally much more well-conditioned than in the steady state case. To see this, consider the matrix S_C given by (4.2.2). This matrix is close to being a (block) diagonal matrix, for if the integral over R in (4.2.2) were evaluated by a numerical method which uses the nodes as the integration points, $(S_C)_{ij}$ would be zero for $i \neq j$. Although the numerical integration method actually used does not use the nodes, a reasonably accurate scheme can be constructed using only nodal points, so the off-diagonal elements must be small relative to the diagonal elements. Now as $dt \to 0$, for a fixed triangulation, G (see 4.1.6) approximates $-S_C/dt_n$ and so $D_G^{-1}G$ is close to the identity. This indicates that, for sufficiently small time step sizes, $D_G^{-1}G$ is well conditioned, with condition number much less than for J itself, and therefore the convergence rate should be much faster.

To compare the relative efficiencies of the band and Lanczos solvers on time dependent problems, the following nonsymmetric test problem was used:

$$u_t = \nabla^2 u + u_x + 1 \qquad \text{in the unit square}$$

with $\qquad u = 1 \qquad$ on the boundary

and $\qquad u = 1 \qquad$ at $t=0$.

The results shown in Tables 4.3.1 and 4.3.2 are from a Data General MV10000, using double precision.

Table 4.3.1

True Newton Iteration (Jacobian re-assembled)

			CPU time (seconds) per Newton iteration		Matrix and vector
quadratic triangles	N	DT	Lanczos (iters.)	Band Solver	assembly
100	180	0.1	3.90 (32)	1.53	2.53
		0.01	2.70 (22)	1.53	2.53
		0.001	1.15 (9)	1.53	2.53
		0.0001	1.27 (10)	1.53	2.53
500	948	0.1	35.16 (55)	42.32	13.46
		0.01	24.00 (38)	42.32	13.46
		0.001	9.94 (15)	42.32	13.46
		0.0001	4.82 (7)	42.32	13.46
2500	4873	0.1	318.63 (101)	1234.76	76.59
		0.01	236.60 (74)	1234.76	76.59
		0.001	101.54 (32)	1234.76	76.59
		0.0001	37.22 (12)	1234.76	76.59

Table 4.3.2

Pseudo-Newton Iteration (Old Jacobian Used)

			CPU time (seconds) per Newton iteration		Vector
quadratic triangles	N	DT	Lanczos (iters.)	Band Solver	assembly
100	180	0.1	3.90 (32)	0.53	0.69
		0.01	2.70 (22)	0.53	0.69
		0.001	1.15 (9)	0.53	0.69
		0.0001	1.27 (10)	0.53	0.69
500	948	0.1	35.16 (55)	6.38	3.37
		0.01	24.00 (38)	6.38	3.37
		0.001	9.94 (15)	6.38	3.37
		0.0001	4.82 (7)	6.38	3.37
2500	4873	0.1	318.63 (101)	62.49	24.38
		0.01	236.60 (74)	62.49	24.38
		0.001	101.54 (32)	62.49	24.38
		0.0001	37.22 (12)	62.49	24.38

When the Jacobian is updated, and the band solver has to re-decompose it
(Table 4.3.1) the Lanczos method has a decisive edge and the more triangles
used, and the smaller the time step, the greater is that edge. If the
matrix is updated very rarely, Table 4.3.2 becomes more significant, however.
When an already decomposed Jacobian is used, the band solver only does
$O(N^{1.5})$ work and the Lanczos method does not gain the advantage until the
time step is very small. See Problem 4.5 for a further analysis of the
relative merits of the two methods, for time-dependent problems. Notice
that the band solver cost behavior is very close to that predicted by the
theory, namely $O(N^2)$ for a true Newton iteration and $O(N^{1.5})$ when an old
Jacobian is used, and that the Lanczos cost behavior is better, for each
time step, than $O(N^{1.5})$.

It should be noted also that the data in these tables are based on the first
(true or pseudo) Newton iteration on a time step. For subsequent iterations
the band method will require the same amount of time, but the Lanczos method
should require less time each Newton iteration, as better and better starting
values are available. This same reasoning applies to nonlinear steady state
problems, giving Lanczos an additional edge there also.

4.4 **Exercises** (* = requires use of PDE/PROTRAN)

4.1 Show that S_C given by (4.2.2) is positive definite and that J given by
(2.1.9) is negative definite, for the problem:

$$c(x,y,t)u_t = \nabla^T(p(x,y,t) \nabla u) - q(x,y,t)u + f(x,y,t) \quad \text{in R}$$

$$u = fb(x,y,t) \quad \text{on } \partial R_1$$

$$p(\partial u/\partial n) = -s(x,y,t)u + gb(x,y,t) \quad \text{on } \partial R_2$$

if c and p are positive and q and s are nonnegative (assume also that
∂R_1 is not empty). Conclude that the eigenvalues of $J\mathbf{w}= \lambda S_C\mathbf{w}$ are all
real and negative.

4.2 Show that if $\mathbf{u}(t)$ is the solution to the linear ODE system:

$$\mathbf{u}_t = C(t)\mathbf{u} + \mathbf{b}(t)$$

$$\mathbf{u}(0) = \mathbf{u_0}$$

and $\mathbf{y}(t_n)$ is the finite difference "Crank-Nicolson" approximation with
constant time step,

$$(\mathbf{y}(t_{n+1})-\mathbf{y}(t_n))/dt = C(t_n+0.5dt)(\mathbf{y}(t_{n+1})+\mathbf{y}(t_n))/2 + \mathbf{b}(t_n+0.5dt)$$

$$\mathbf{y}(0) = \mathbf{u_0}$$

then the error satisfies:

$$\mathbf{e}(t_n) = \mathbf{y}(t_n)-\mathbf{u}(t_n) = \mathbf{p}(t_n)dt^2 + O(dt^4)$$

(Hint: First show that the error satisfies:

$$(\mathbf{e}(t_{n+1})-\mathbf{e}(t_n))/dt = C(t_n+0.5dt)(\mathbf{e}(t_{n+1})+\mathbf{e}(t_n))/2 + \mathbf{T}(t_n+0.5dt)$$

$$\mathbf{e}(0) = \mathbf{0}$$

where $\mathbf{T}(t)=\mathbf{q}(t)dt^2 + O(dt^4)$ is the truncation error. Then if $\mathbf{p}(t)$
satisfies the ODE system:

$$\mathbf{p}_t = C(t)\mathbf{p} + \mathbf{q}$$

$$\mathbf{p}(0) = \mathbf{0}$$

show that $\mathbf{e}(t_n)-dt^2 \mathbf{p}(t_n) = O(dt^4)$.)

4.3 Suppose the nonlinear ODE:

$$du/dt = f(t,u)$$

is solved using the "Crank-Nicolson" scheme:

$$(y(t_{n+1})-y(t_n))/dt = f(t_n+0.5dt \, , \, (y(t_{n+1})+y(t_n))/2)$$

Show that if a single Newton iteration is done each time step to solve this nonlinear equation for $y(t_{n+1})$, using an initial guess of $y(t_n)$, the resulting difference equation still has truncation error $O(dt^2)$. This indicates that one true Newton iteration per time step for the system (4.1.4) is sufficient, which is what is done if neither "NOUPDATE" nor "AUTOUPDATE" is specified.

*4.4 Use both the Crank-Nicolson and backward difference methods to solve:

$$u_t = \nabla^2 u \qquad \text{in the rectangle } (-1,1) \text{ X } (0,1)$$

with

$$\partial u/\partial n = 0 \qquad \text{on the boundary}$$

and

$$u = \begin{cases} 1 & \text{if x is positive} \\ 0 & \text{if x is negative} \end{cases} \qquad \text{at } t=0$$

Use a triangulation of 8 triangles as in Problem 2.9, and a time step of 0.1. Notice the oscillation in the solution when the Crank-Nicolson method is used, due to the unsmooth initial conditions.

4.5 According to Tables 4.3.1-2 when, say, 2500 quadratic triangles and a time step of 0.01 are used, the Lanczos method is much faster for a true Newton iteration, whereas the band solver is faster when the Jacobian is not updated. Thus if the Jacobian has to be updated often, the Lanczos method has the advantage, while if it is rarely updated, the band solver will win. Find the break-even point, that is, find m such that if the Jacobian is updated every mth Newton iteration, the two methods are about even. Find the break-even point for the other time step sizes, for 2500 triangles.

*4.6 Solve the backwards heat equation:

$$u_t = -\nabla^2 u \qquad \text{in the unit square}$$

$$u = 0 \qquad \text{on } \partial R$$

$$u = x(1-x)y(1-y) \qquad \text{when } t=0$$

Use either time discretization method, with a couple of different triangulations and time step sizes. Is the method stable? Why or why not?

References

1. G. E. Forsythe and W. R. Wasow, _Finite Difference Methods for Partial Differential Equations_, John Wiley and Sons, New York (1960).

2. A. R. Mitchell and D. F. Griffiths, _The Finite Difference Method in Partial Differential Equations_, John Wiley and Sons, New York, (1980).

CHAPTER 5. HYPERBOLIC PROBLEMS

While PDE/PROTRAN is competitive in speed and accuracy with other software
for the solution of elliptic and parabolic problems, it is certainly not
ideally designed to handle hyperbolic problems. Nevertheless, because of
its ease of use and flexibility, it is expected that it will be occasionally
used for hyperbolic problems. The results of this chapter will show, in
fact, that for both first and second order hyperbolic problems, PDE/PROTRAN
is a useful tool provided the solutions being approximated are smooth, and
is probably as robust as other general purpose PDE packages for these
problems.

Unfortunately, a property of hyperbolic equations is that, unlike elliptic
and parabolic equations, they propagate discontinuities and other rough
behavior from the initial and boundary data into the interior of the region,
and thus solutions quite often are not smooth. Worse, the regions of rough
behavior are often mobile, and thus not treatable by a graded, fixed grid.
For such problems only special methods tailored to hyperbolic problems are
suitable, and neither PDE/PROTRAN nor any other general purpose code is
recommended.

5.1 First Order Transport Problems

Consider the first order "transport" problem (cf. the convection problem
1.2.1):

$$u_t = F(x,y,t,u,u_x,u_y) \qquad \text{in R}$$

$$(5.1.1) \qquad An_x + Bn_y = On_x + On_y = \beta E(u-fb(x,y,t)) \qquad \text{on } \partial R$$

$$u = UO(x,y) \qquad \text{at } t=t_0$$

where β is a large number (see Problem 2.10) and $E(x,y,t)$ is a diagonal
matrix whose entries are always either $E_{ii} = 1$, when $u_i = fb_i$ on the boundary,
or $E_{ii} = 0$, when u_i is not specified (so that the equation reduces to 0=0,
or no boundary condition.) Thus on each boundary segment, some, but not
necessarily all, of the unknowns may be specified.

To study the stability of the PDE/PROTRAN approximation and to determine the
boundary conditions appropriate for this problem, it is convenient to
consider the linearized version of (5.1.1), where G=F.UX, H=F.UY and C=F.U:

$$(5.1.2) \qquad u_t = G(x,y,t)u_x + H(x,y,t)u_y + C(x,y,t)u + f(x,y,t) \qquad \text{in R}$$

$$u = UO(x,y) \qquad \text{at } t=t_0$$

where G,H and C are m by m matrices. By definition {1, Section 4.6} this
system is hyperbolic if $n_x G + n_y H = P^{-1}(x,y,t)D(x,y,t)P(x,y,t)$ for any
n_x, n_y, where D is a real diagonal matrix.

To see what boundary conditions are appropriate for (5.1.2), integrate both
sides of the PDE system over R:

$$\iint_R \mathbf{u}_t \; dxdy = \iint_R (G\mathbf{u}_x + H\mathbf{u}_y + C\mathbf{u} + \mathbf{f}) \; dxdy$$

or

$$d/dt \iint_R \mathbf{u} \; dxdy = \int_{\partial R} (n_x G + n_y H)\mathbf{u} \; ds + \iint_R (C\mathbf{u} + \mathbf{f} - G_x\mathbf{u} - H_y\mathbf{u}) \; dxdy$$

Clearly (see Section 1.2) the boundary integral represents the change in $\mathbf{u}$ in the region due to the flux across the boundary, so that the change in $\mathbf{u}$ due to flux across a small piece ds of the boundary during time interval dt is:

$$\Delta\mathbf{u} = (n_x G + n_y H)\mathbf{u} \; dsdt = P^{-1}DP \; \mathbf{u} \; dsdt$$

Making the change of dependent variable $\mathbf{v} = P\mathbf{u}$ gives:

$$\Delta\mathbf{v} = D\mathbf{v} \; dsdt$$

From this it is seen that the net flux of component v_k is inward if D_{kk} is positive on this piece of the boundary, and outward if D_{kk} is negative. Now on that part of the boundary where D_{kk} is negative, and the flux is outward, it seems unreasonable to specify v_k (see Problem 1.4).

In fact {cf.1, Section 4.4} along a boundary segment where q of the m eigenvalues D_{kk} are positive, there should be specified exactly q unknowns. Any q of the variables u_k may be specified, so long as none of the m−q components of $\mathbf{v}$ corresponding to nonpositive D_{kk} happen to be linear combinations of these u_k, and thus implicitly specified.

To investigate the stability of the PDE/PROTRAN Galerkin approximation to the linear problem (5.1.2), two simplifying assumptions will be made. There is no experimental evidence that either is actually necessary for stability, however. First, assume G and H are symmetric (thus $n_x G + n_y H = P^{-1}DP = P^TDP$ for any n_x, n_y and the problem is hyperbolic). Second, assume that on any boundary segment all of the eigenvalues D_{kk} have the same sign. Notice that when m=1, both assumptions are automatically satisfied. Thus on that part of the boundary (∂R_1) where all D_{kk} are positive, $\mathbf{u}$ is given, while on the other part (∂R_2) $A n_x + B n_y = 0$, which is equivalent to supplying no boundary condition.

As in Section 4.2, to prove stability it is necessary only to show that the real parts of all eigenvalues of $J\mathbf{w} = \lambda S_I \mathbf{w}$ are bounded above by a positive constant independent of the triangulation, where J is given by (2.1.9) and S_I by (4.2.2) with C replaced by the identity I.

Now if $J\mathbf{w} = \lambda S_I \mathbf{w}$, where $\mathbf{w}$ has components w_k and $W = \sum w_k \phi_k$:

$$2\mathrm{Re}(\lambda) \iint_R |W|^2 \; dxdy = (\lambda + \lambda^*)\mathbf{w}^* T S_I \mathbf{w} = \mathbf{w}^* T J\mathbf{w} + \mathbf{w}^T J\mathbf{w}^* = \mathbf{w}^* T(J + J^T)\mathbf{w}$$

Using (2.1.11), the right hand side becomes:

$$\iint\limits_{R} \{\mathbf{W}^{*T}(C+C^T)\mathbf{W} + \mathbf{W}^{*T}G\mathbf{W}_x + \mathbf{W}_x^{*T}G\mathbf{W} + \mathbf{W}^{*T}H\mathbf{W}_y + \mathbf{W}_y^{*T}H\mathbf{W}\} \, dxdy$$

$$= \iint\limits_{R} \mathbf{W}^{*T}(C+C^T-G_x-H_y)\mathbf{W} \, dxdy + \int\limits_{\partial R_2} \mathbf{W}^{*T}(n_xG+n_yH)\mathbf{W} \, ds$$

$$\leq M \iint\limits_{R} |\mathbf{W}|^2 \, dxdy + \int\limits_{\partial R_2} \mathbf{W}^{*T}P^TDP\mathbf{W} \, ds$$

where M is the largest eigenvalue of the symmetric matrix $C+C^T-G_x-H_y$ in R, and the fact that all of the ϕ_k, and thus $\mathbf{W}$, are zero on ∂R_1 has been used. The remaining boundary integral is nonpositive, since $\mathbf{W}^{*T}P^TDP\mathbf{W} = (P\mathbf{W})^{*T}D(P\mathbf{W})$ and the D_{kk} are all nonpositive on ∂R_2. Finally,

$$2 \, \mathrm{Re}(\lambda) \leq M$$

and both the Crank-Nicolson and backward difference methods are stable.

Now consider the convection problem (1.2.1):

$$u_t = -Gu_x - Hu_y + f(x,y,t,u)$$

It is physically evident that if discontinuities or other roughness exist in the initial concentration distribution or boundary values, they will simply be propagated by the "wind" or "current" vector (G,H), with no smoothing. This is characteristic of hyperbolic problems.

If, however, a small amount of "artificial diffusion" is added:

$$(5.1.3) \quad u_t = (0.5|G|h \, u_x)_x + (0.5|H|h \, u_y)_y - Gu_x - Hu_y + f(x,y,t,u)$$

where h is some approximate measure of the element size near (x,y), it is clear physically that the diffusion will smooth out the rough edges in the solution. This technique is often necessary to stabilize the calculations for transport problems. Mathematically, the second order terms cause the stability eigenvalues μ_k to go to zero faster with k and dampen the high frequency noise more rapidly (see Problem 5.1).

As the triangulation is refined the diffusion terms go to zero, and although the amount of diffusion added above is somewhat arbitrary, it is motivated by the following analysis. Consider the finite difference approximation to (5.1.3) analogous to the backward difference method, using centered spatial differences, with a step size of $\Delta x = \Delta y = h$. Assume G,H and h are constant, for simplicity. Then ($u_{ij}^n = u(x_i,y_j,t_n)$):

$$(u_{ij}^{n+1} - u_{ij}^n)/dt = 0.5|G|h(u_{i+1,j} - 2u_{ij} + u_{i-1,j})/h^2$$

$$+0.5|H|h(u_{i,j+1} - 2u_{ij} + u_{i,j-1})/h^2$$

$$-G(u_{i+1,j} - u_{i-1,j})/(2h) - H(u_{i,j+1} - u_{i,j-1})/(2h) + f(x_i,y_j,t_{n+1},u_{ij})$$

where all u terms on the right hand side are evaluated at level n+1.

If G is positive, the term multiplying -G becomes $(u_{ij}-u_{i-1,j})/h$ while if it is negative, the term is $(u_{i+1,j}-u_{ij})/h$, and similarly for the coefficient of -H. Since the wind velocity is (G,H), this means that adding the artificial diffusion in (5.1.3) converts central differences into "upwind" differences, that is, differences which involve u_{ij} and its neighbors in the upwind directions. For the explicit finite difference method, this makes an unconditionally unstable method (central differences) into a conditionally stable one (upwind differences). The implicit method is stable even with central differences, but as shown in Problem 5.1, upwinding causes the high frequency noise to be damped out more rapidly. Problem 5.1 also shows that adding exactly the amount of artificial diffusion, or viscosity, suggested in (5.1.3) is not critical.

The situation is much more complex with the finite element method, but it is comforting to know that at least in the finite difference case, the addition of a small amount of artificial viscosity is well justified theoretically, even though it changes the nature of the problem from hyperbolic to parabolic. In the finite element case it is well justified experimentally.

As an example, consider the following nonlinear shallow water flow problem, taken from {2}. The region involved is one-dimensional, so PDE/PROTRAN treats it as a long thin rectangle (cf. Problem 3.2d). The equations are:

$$u_t = -uu_x - 980(v_x+H_x) \qquad -400 \le x \le 400$$

(5.1.4)

$$v_t = -uv_x - vu_x$$

and boundary conditions:

$$u(-400,t) = 40$$
$$v(400,t) = 20$$

and initial conditions:

$$u(x,0) = 40$$
$$v(x,0) = 20-H(x)$$

Here u represents the horizontal velocity and v represents the depth of a fluid flowing over a surface whose height is defined by $H(x) = \max(0,10-10(x/40)^2)$. Thus there is a barrier centered at x=0.

For this nonlinear problem, the matrix corresponding to G in (5.1.2) is the Jacobian:

$$F.UX = \begin{bmatrix} -u & -980 \\ -v & -u \end{bmatrix}$$

while the matrix corresponding to H is F.UY=0. The eigenvalues of $n_xG+n_yH = n_xG$ are $\lambda = n_x(-u+(980v)^{0.5})$, $\lambda = n_x(-u-(980v)^{0.5})$, so as long as the depth (v) is positive the problem is hyperbolic. Since near each end u and v are very nearly 40 and 20, respectively, at all times, the two eigenvalues have opposite sign at both ends (e.g when $n_x=1$, the eigenvalues are 100 and

-180).   Thus since it can be verified that none of the eigenvectors are
parallel to u or v (Problem 5.4), at each end either u or v can be specified.
The above boundary conditions specify u at the upstream end (x=-400) and v
at the downstream end.

The PDE/PROTRAN input is shown in Table 5.1.1.   At about t=0.7 a shock wave,
leading to steep gradients in u and v, begins to develop and PDE/PROTRAN is
unable to solve the problem beyond this point.   Until then, while the
solution is relatively smooth, the results are good (see Figure 5.1.1).
Similar difficulties were encountered for this problem using IMSL subroutine
DPDES {3} and Sincovec and Madsen's PDEONE {2}, both of which use a method
of lines approach with an adaptive Gear's method to solve the ODE system
(4.1.3), and thus are very robust routines.

When an artificial diffusion term was added, PDE/PROTRAN, DPDES and PDEONE
were all able to continue past where the shock wave would have developed.
The linearized problem, where u and v are replaced by 40 and 20 in (5.1.4),
was integrated successfully by PDE/PROTRAN to its steady state condition,
without using artificial viscosity, indicating that it is the nonlinearity
of the problem which allows the shock wave to develop.

Table 5.1.1

```
C     HYPERBOLIC SHALLOW WATER FLOW PROBLEM
C
$     PDE2D
      UNKNOWNS = (U,V)
      C = (1,1)
      F = (-U*UX-980.*(VX+HX(X))  ,  -U*VX-V*UX )
C   INITIAL CONDITIONS
      UO = (40,20-H(X))
C   BOUNDARY CONDITIONS
      GB = (1, 1.E8*(U-40.) , 0 )
    *     (2, 0  , 1.E8*(V-20) )
C   GRID MOST DENSE BETWEEN -60 AND 60
      XGRID = -400,-200,-100,-80,-70,-60,-55,-50,-45,-40,-35,-30,-25,-20,
    *  -15,-10,-5,0,5,10,15,20,25,30,35,40,45,50,55,60,70,80,100,200,400
      YGRID = 0,1
      ARCS = (1,2) (3,4)
      SAVEFILE = PLOT
      OUTFREQUENCY = 10
      DT = 0.01
      TF = 0.6
      AUTOUPDATE
C   OUTPUT ALONG SINGLE HORIZONTAL LINE Y=0
      GRIDPOINTS = (101,1)
      GRIDLIMITS = (-200,200) (0,0)
$     PLOTVECTOR
      SET = 6
      VECTOR = (0,U)
      TITLE = 'VELOCITY AT TIME=0.6'
$     END
```

```
$       FORTRAN
        REAL FUNCTION H(X)
        H = AMAX1(0.0,10.-10.*(X/40)**2)
        RETURN
        END
        REAL FUNCTION HX(X)
        HX = 0.0
        IF (ABS(X).LE.40.0) HX = -X/80.0
$       END
```

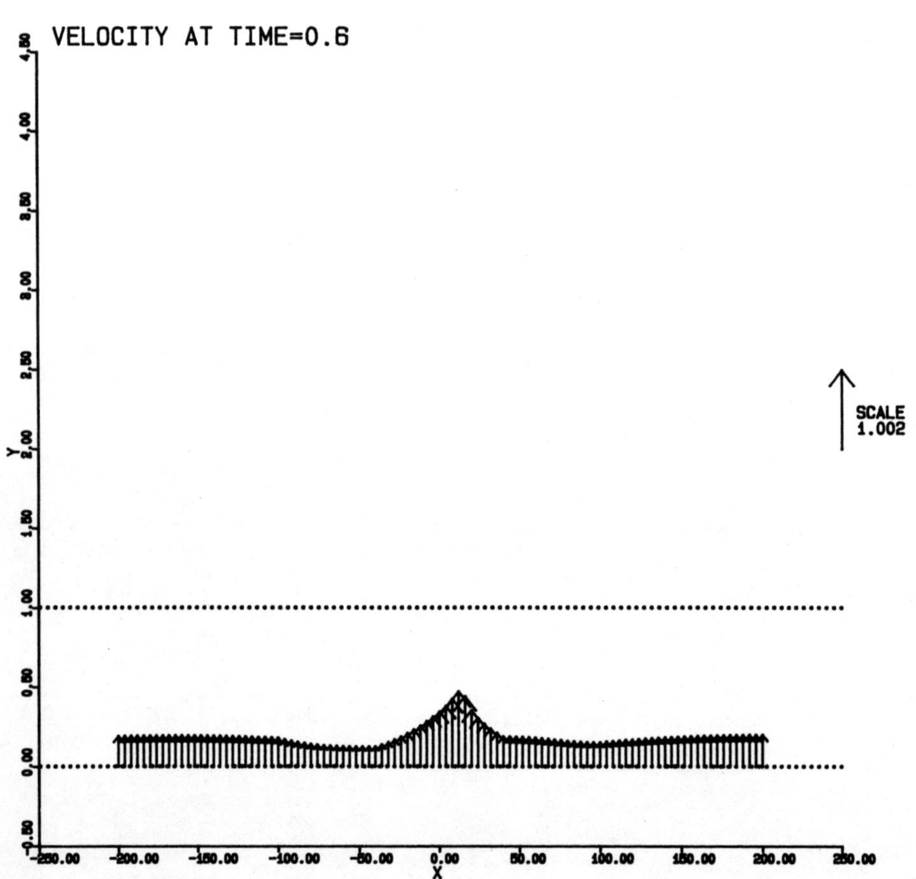

Figure 5.1.1 Velocity Distribution for Shallow Water Flow Problem

100

5.2 Second Order Wave Problems

Consider now the second order wave problem (cf. 1.3.1 and 1.3.2):

$$C(x,y,t,\mathbf{u})\mathbf{u}_{tt} = \mathbf{A}_x(x,y,t,\mathbf{u},\mathbf{u}_x,\mathbf{u}_y) + \mathbf{B}_y(x,y,t,\mathbf{u},\mathbf{u}_x,\mathbf{u}_y) + \mathbf{F}(x,y,t,\mathbf{u},\mathbf{u}_x,\mathbf{u}_y) \text{ in } R$$

$$\mathbf{u} = \mathbf{FB}(x,y,t) \qquad \text{on } \partial R_1 \tag{5.2.1}$$

$$\mathbf{A}n_x + \mathbf{B}n_y = \mathbf{GB}(x,y,t,\mathbf{u}) \qquad \text{on } \partial R_2$$

$$\begin{aligned} \mathbf{u} &= \mathbf{UO}(x,y) \\ \mathbf{u}_t &= \mathbf{U1}(x,y) \end{aligned} \qquad \text{at } t{=}t_0$$

This can be reduced to a system of 2m (m=dimension of **u**) PDEs of the form (4.1.1) by introducing the auxiliary variable $\mathbf{v}{=}\mathbf{u}_t$. Then (5.2.1) becomes:

$$\mathbf{u}_t = \mathbf{v} \qquad \text{in } R$$

$$C(x,y,t,\mathbf{u})\mathbf{v}_t = \mathbf{A}_x(x,y,t,\mathbf{u},\mathbf{u}_x,\mathbf{u}_y) + \mathbf{B}_y(x,y,t,\mathbf{u},\mathbf{u}_x,\mathbf{u}_y) + \mathbf{F}(x,y,t,\mathbf{u},\mathbf{u}_x,\mathbf{u}_y)$$

$$\begin{aligned} \mathbf{u} &= \mathbf{FB}(x,y,t) \\ \mathbf{v} &= \mathbf{FB}_t(x,y,t) \end{aligned} \qquad \text{on } \partial R_1$$

$$\begin{aligned} \mathbf{O}n_x + \mathbf{O}n_y &= \mathbf{0} \\ \mathbf{A}n_x + \mathbf{B}n_y &= \mathbf{GB}(x,y,t,\mathbf{u}) \end{aligned} \qquad \text{on } \partial R_2 \tag{5.2.2}$$

$$\begin{aligned} \mathbf{u} &= \mathbf{UO}(x,y) \\ \mathbf{v} &= \mathbf{U1}(x,y) \end{aligned} \qquad \text{at } t{=}t_0$$

Notice that the **A** and **B** corresponding to the first equation are **0**, so that the first boundary condition on ∂R_2 is appropriate, and is equivalent to supplying no boundary condition at all.

Again, as shown in Section 4.2, to prove that the two PDE/PROTRAN time discretization methods are stable for the linearized version of (5.2.2), it is necessary to show that the real parts of the eigenvalues of $J\mathbf{w} = \omega S\mathbf{w}$ are bounded above by a constant independent of the spatial grid. But by comparing (5.2.2) with (4.1.1), it is seen that the 2mN by 2mN matrices corresponding to J and S for this system of 2m PDEs can be written, respectively, as:

$$\begin{bmatrix} 0 & S_I \\ J & 0 \end{bmatrix} \qquad \text{and} \qquad \begin{bmatrix} S_I & 0 \\ 0 & S_C \end{bmatrix}$$

where the mN by mN matrices J, S_I and S_C are given, as before, by (2.1.9) and (4.2.2). Thus the eigenvalues ω of interest are found by setting:

$$0 = \det \begin{bmatrix} -\omega S_I & S_I \\ J & -\omega S_C \end{bmatrix}$$

Multiplying the above matrix by:

$$\begin{bmatrix} I & 0 \\ JS_I^{-1}/\omega & I \end{bmatrix}$$

will not change the determinant, since the determinant of the multiplying matrix is one, and the determinant of a product is the product of the determinants. Thus:

$$0 = \det \begin{bmatrix} -\omega S_I & S_I \\ 0 & -\omega S_C + J/\omega \end{bmatrix}$$

$$= \det(-\omega S_I) \, \det(-\omega S_C + J/\omega)$$

Since S_I is positive definite, either $\omega=0$ or ω is an eigenvalue of $J\mathbf{w} = \omega^2 S_C\mathbf{w}$. This means that the square roots of the eigenvalues of the problem $J\mathbf{w} = \lambda S_C\mathbf{w}$ must have real parts bounded above by a constant. Since λ will have two square roots with real parts opposite in sign and equal in absolute value, this means that the condition for stability is:

$$| \mathrm{Re}(\lambda^{0.5}) | \leq M$$

This stability condition is certainly fulfilled for those problems where S_C is positive definite and J is negative definite (e.g. Problem 4.1), since then the λ are all real and negative, and thus $\mathrm{Re}(\lambda^{0.5})=0$. In this case there is no threshold limit (see Section 4.2) for dt_n, since M can be taken arbitrarily small. Notice that for such problems, all the stability eigenvalues for the Crank-Nicolson method have absolute value equal to one (see 4.2.3 and recall that λ must be replaced by ω). Thus high frequency noise in the initial data, i.e. components along $\mathbf{w}_k$ where $|\lambda_k|$ is large, are not damped at all. This mirrors the behavior of the PDE itself, since rough initial data do not become smooth with time, if the equation is hyperbolic.

To relate the stability condition to the eigenvalues themselves, rather than their square roots, consider the conditions:

$$\mathrm{Re}(\lambda) \leq M \qquad (M, K_1, K_2 \text{ positive})$$

(5.2.3)

$$|\mathrm{Im}(\lambda)|^2 \leq K_1 - K_2\mathrm{Re}(\lambda)$$

It will be shown that these imply that $|\mathrm{Re}(\lambda^{0.5})|$ is bounded.

Geometrically this limits the eigenvalues to the shaded area of Figure 5.2.1 so that they can become infinite only in the general direction of the negative real axis.

Now when $|\text{Re}(\lambda)| \leq M$, $|\text{Re}(\lambda^{0.5})|$
is certainly bounded (see Figure
5.2.1). So it may be assumed
that $\text{Re}(\lambda) \leq -M$. Now if
$\lambda = |\lambda|\cos\theta + i|\lambda|\sin\theta$ then
$\lambda^{0.5} = |\lambda|^{0.5}\cos(\theta/2)+i|\lambda|^{0.5}\sin(\theta/2)$
(and its negative) and so:

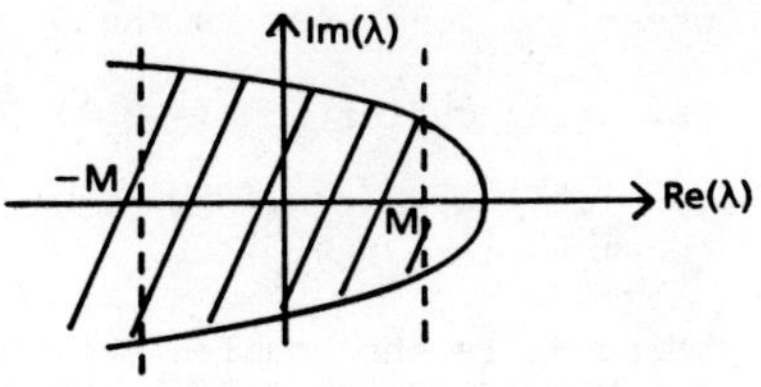

Figure 5.2.1

$$2|\text{Re}(\lambda^{0.5})|^2 = 2|\lambda|\cos^2(\theta/2) = |\lambda|(1+\cos\theta) = |\lambda|+\text{Re}(\lambda) =$$

$$\frac{|\lambda|^2-(\text{Re}(\lambda))^2}{|\lambda| - \text{Re}(\lambda)} = \frac{(\text{Im}(\lambda))^2}{|\lambda|+|\text{Re}(\lambda)|} \leq \frac{(\text{Im}(\lambda))^2}{2|\text{Re}(\lambda)|} \leq \frac{K_1-K_2\text{Re}(\lambda)}{2|\text{Re}(\lambda)|}$$

$$\leq K_1/(2M) + K_2/2$$

and so $|\text{Re}(\lambda^{0.5})|$ is bounded, as claimed.

Conditions (5.2.3) are satisfied for any linear problem of the form (5.2.1)
such that the elements of C are positive and such that the 2m by 2m matrix:

$$\begin{bmatrix} A.UX & A.UY \\ B.UX & B.UY \end{bmatrix}$$

is positive definite. Although not necessary, to simplify the proof it will
be assumed that **GB** is not a function of **u**.

It has already been shown, under assumptions satisfied here, that $\text{Re}(\lambda)$ is
bounded above (see 4.2.5). Thus, $\text{Im}(\lambda)$ must now be bounded, in the form
(5.2.3). But:

$$2i\,\text{Im}(\lambda)\,\mathbf{w}^{*T}S_C\mathbf{w} = (\lambda-\lambda^*)\mathbf{w}^{*T}S_C\mathbf{w} = \mathbf{w}^{*T}J\mathbf{w} - \mathbf{w}^TJ\mathbf{w}^* = \mathbf{w}^{*T}(J-J^T)\mathbf{w}$$

Using (2.1.11) the right hand side can be written (recall that $\mathbf{W}=\sum\mathbf{w_k}\phi_k$):

$$\iint\limits_{R} (\mathbf{w}^*,\mathbf{w_x}^*,\mathbf{w_y}^*)^T \begin{bmatrix} F.U-F.U^T & F.UX+A.U^T & F.UY+B.U^T \\ -A.U-F.UX^T & 0 & 0 \\ -B.U-F.UY^T & 0 & 0 \end{bmatrix} \begin{bmatrix} \mathbf{W} \\ \mathbf{W_x} \\ \mathbf{W_y} \end{bmatrix} dxdy$$

So now

$$2|\text{Im}(\lambda)|\,\mathbf{w}^{*T}S_C\mathbf{w} \leq \iint \{2||F.UX+A.U^T||\,||\mathbf{W}||\,||\mathbf{W_x}||$$

$$+ ||F.U-F.U^T||\,||\mathbf{W}||^2 + 2||F.UY+B.U^T||\,||\mathbf{W}||\,||\mathbf{W_y}||\} dxdy$$

where the norm used is the L_2 norm, for m-vectors and m by m matrices.

Now using definition (4.2.2) for S_C:

$$\mathbf{w}^{*T}S_C\mathbf{w} = \iint_R \mathbf{w}^{*T}C\mathbf{w}\ dxdy \geq \iint_R \sigma_c\ ||\mathbf{w}||^2\ dxdy$$

where σ_c is the smallest eigenvalue of the positive definite diagonal matrix C. Now if the following definitions are made:

$$\alpha_1 = \min_R \sigma_c \qquad \alpha_2 = \max(\max_R ||F.UX+A.U^T||, \max_R ||F.UY+B.U^T||)$$
$$\alpha_3 = \max_R ||F.U-F.U^T||$$

then:

$$2|\text{Im}(\lambda)|\alpha_1 \iint ||\mathbf{w}||^2\ dxdy$$

$$\leq \alpha_3\iint ||\mathbf{w}||^2\ dxdy + 2\alpha_2\iint ||\mathbf{w}||(||\mathbf{w}_x||+||\mathbf{w}_y||)dxdy$$

$$\leq \alpha_3\iint ||\mathbf{w}||^2\ dxdy + 2\alpha_2 \{\iint ||\mathbf{w}||^2dxdy\}^{0.5} \{\iint (||\mathbf{w}_x||+||\mathbf{w}_y||)^2dxdy\}^{0.5}$$

If z is defined by:

$$z^2 = (\iint ||\mathbf{w}_x||^2 + ||\mathbf{w}_y||^2\ dxdy)\ /\ (\iint ||\mathbf{w}||^2\ dxdy)$$

Then:

$$|\text{Im}(\lambda)| \leq \gamma_1 z + \gamma_2$$

But it was shown in (4.2.5) that:

$$(z-\beta_1)^2 \leq \beta_2 - \beta_3\text{Re}(\lambda)$$

where z is the same as in this section, and the β_i and γ_i are all nonnegative constants. Now:

$$|\text{Im}(\lambda)|^2 \leq (\gamma_1(z-\beta_1) + \gamma_2 + \gamma_1\beta_1)^2$$

$$\leq (\gamma_1(\beta_2-\beta_3\text{Re}(\lambda)+1)^{0.5} + \gamma_2 + \gamma_1\beta_1)^2$$

$$\leq (\beta_2-\beta_3\text{Re}(\lambda)+1)\ (\gamma_1 + (\gamma_2+\gamma_1\beta_1)/(\beta_2-\beta_3\text{Re}(\lambda)+1)^{0.5})^2$$

$$\leq (\beta_2-\beta_3\text{Re}(\lambda)+1)\ (\gamma_1 + \gamma_2 + \gamma_1\beta_1)^2$$

which verifies that the second part of (5.2.3) holds.

As mentioned earlier, PDE/PROTRAN is not expected to be competitive with special purpose programs for hyperbolic problems. One reason for this is the fact that explicit finite difference methods are generally used on hyperbolic problems, eliminating the need to solve a large linear system each time step. (Although explicit methods are only conditionally stable for both types of problems, the conditions require dt to be unreasonably small for parabolic problems, but not for hyperbolic problems.) The fact that the PDE/PROTRAN band solvers may factor the Jacobian matrix very rarely, or never, may partially offset this advantage.

As an example, consider the time dependent elasticity problem (1.3.2) with plane strain (1.1.5) constitutive equations, in the notched plate of Figure 5.2.1. Material parameters are taken to be $\rho=1$, $E=11 \times 10^6$, $\nu = 0.2$. The base is clamped and on the top a downward force is suddenly applied at t=0. On the rest of the boundary the forces are zero.

Initially, immediately before the application of the load, the displacements and their velocities $(v_1, v_2) = ((u_1)_t, (u_2)_t)$ are zero. The PDE/PROTRAN input to solve this problem in the form (5.2.2) is shown in Table 5.2.1, and the displacement fields at $t=5 \times 10^{-4}$ and 15×10^{-4} are shown in Figures 5.2.1 and 5.2.2.

For many hyperbolic wave problems of the form (5.2.1), the nonhomogeneous forcing terms are periodic in time (cf. 1.4.1). After the effect of the initial conditions has been damped out (cf. Problem 1.5), the solution will also be periodic with the same period P. If the solution is expanded in a Fourier series of the form:

$$\mathbf{u}(x,y,t) = \sum_{k=1}^{\infty} \mathbf{a_k}(x,y) \sin(2\pi kt/P) + \sum_{k=0}^{\infty} \mathbf{b_k}(x,y) \cos(2\pi kt/P)$$

and similarly for the forcing terms, there will result a series of steady state PDEs for the unknown coefficients $\mathbf{a_k}(x,y)$ and $\mathbf{b_k}(x,y)$ (cf. 1.4.2). See the User's Manual for an example showing how the flexibility of PDE/PROTRAN, which allows PDE/PROTRAN procedures to be intermixed with FORTRAN, allows these systems to be solved successively and the series summed, all in one program.

Table 5.2.1

```
C      TIME DEPENDENT ELASTICITY PROBLEM
C
$      PDE2D
       UNKNOWNS = (U1,U2,V1,V2)
       NTRIANGLES = 50
       FRONTAL
       C = (1,1,1,1)
       A = (0,0, 12.22E6*U1X+3.05E6*U2Y , 4.58E6*(U1Y+U2X) )
       B = (0,0, 4.58E6*(U1Y+U2X) , 12.22E6*U2Y+3.05E6*U1X )
       F = (V1,V2,0,0)
C   PARAMETRIC EQUATIONS OF CURVED NOTCH
       CURVES = (1,SIN(3.14159*S),-COS(3.14159*S))
C   UNIT DOWNWARD FORCE ON TOP
       GB = (3,0,0,0,-1)
C   UPDATING OF JACOBIAN UNNECESSARY
       NOUPDATE
       TO = 0
       DT = 0.00005
       TF = 0.0015
       GRIDPOINTS = (9,17)
       OUTFREQUENCY = 10
       CRANKNICOLSON
       SAVEFILE = PLOT
C   INITIAL TRIANGULATION DEFINITION
       VERTICES = (0,-4) (4,-4) (2,-2) (0,-1) (1,0) (4,0) (0,1) (2,2)
     * (0,4) (4,4) (0.707106,-0.707106) (0.707106,0.707106)
       TRIANGLES = (1,2,3,-1) (2,6,3,2) (6,5,3,0) (5,11,3,1) (11,4,3,1)
     * (4,1,3,5) (5,6,8,0) (6,10,8,2) (10,9,8,3) (9,7,8,4) (7,12,8,1)
     * (12,5,8,1)
$      PLOTVECTOR
       SET = 1
       VECTOR = (U1,U2)
       NODISTORT
       TITLE = 'DISPLACEMENTS AT T=0.0005'
$      PLOTVECTOR
       SET = 3
       VECTOR = (U1,U2)
       NODISTORT
       TITLE = 'DISPLACEMENTS AT T=0.0015'
$      END
```

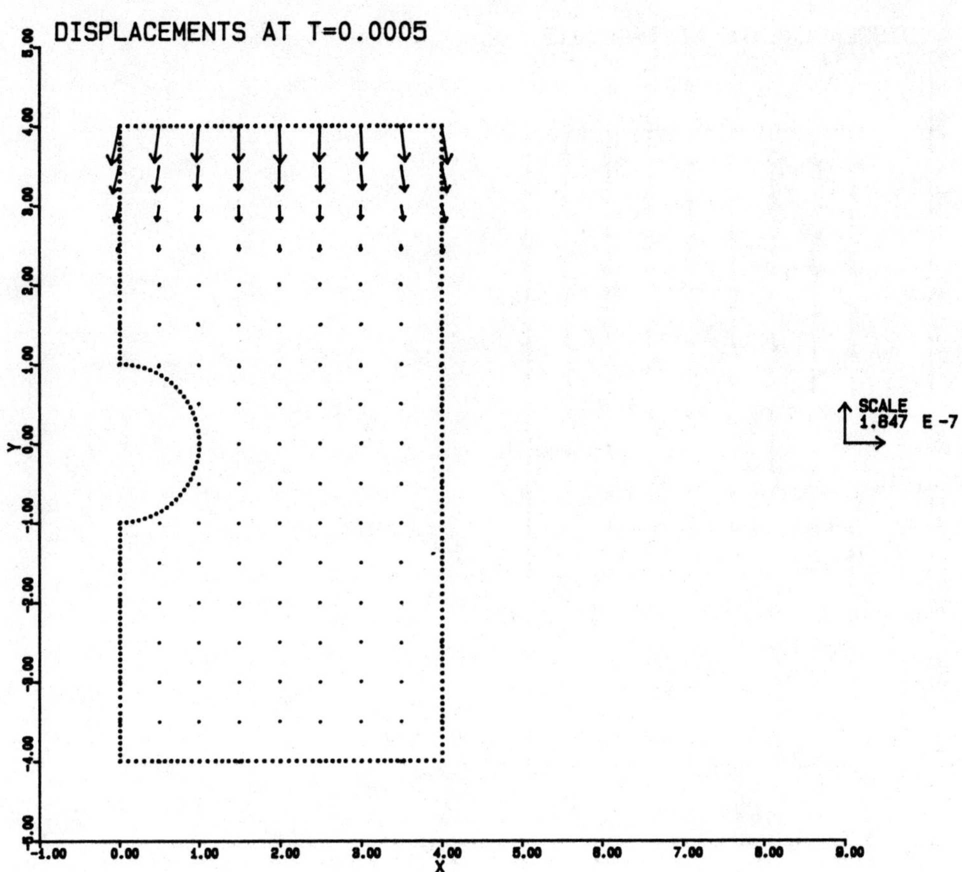

Figure 5.2.1   Displacements for Time Dependent Elasticity Problem (T=0.0005)

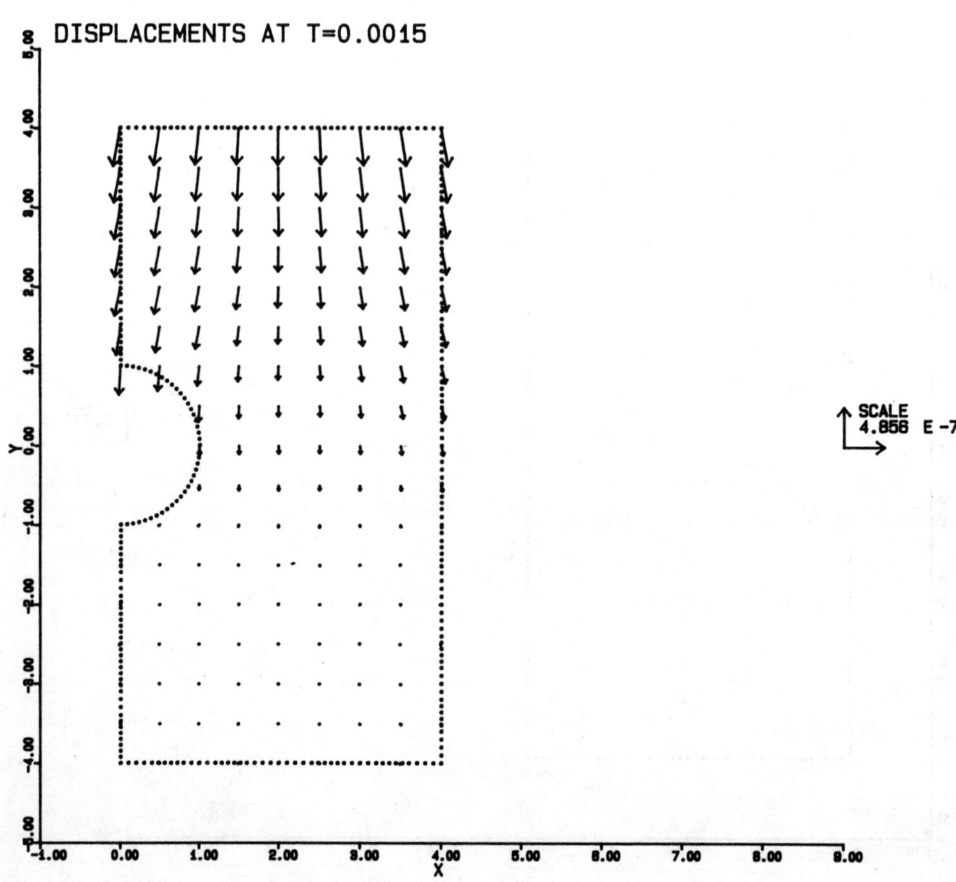

Figure 5.2.2   Displacements for Time Dependent Elasticity Problem (T=0.0015)

108

5.3 **Exercises** (* = requires use of PDE/PROTRAN)

5.1 Consider the linear PDE:

$$u_t = au_{xx} + bu_x \qquad 0 \leq x \leq \pi \ , \ 0 \leq t \leq t_f \ , \ a > 0$$

The Crank-Nicolson ($\alpha=0.5$) and backward difference ($\alpha=1$) finite difference approximations are defined by:

$$(u_j^{n+1} - u_j^n)/dt = a\alpha\Delta^2 u^{n+1} + a(1-\alpha)\Delta^2 u^n + b\alpha\Delta u^{n+1} + b(1-\alpha)\Delta u^n$$

where

$$\Delta^2 u = (u_{j+1} - 2u_j + u_{j-1})/dx^2$$

$$\Delta u = (u_{j+1} - u_{j-1})/(2dx)$$

The Fourier stability analysis technique consists of expanding $u(x_j, t_n)$ as:

$$u_j^n = \sum_{k=1}^{N} a_k(t_n) \exp(ikx_j) \qquad N = \pi/dx, \ x_j = jdx, \ t_n = ndt$$

and substituting this expression into the finite difference equations. For each k, an equation of the form $a_k(t_{n+1}) = \mu_k a_k(t_n)$ results, where the μ_k are the stability eigenvalues. Thus:

$$u_j^n = \sum (\mu_k)^n a_k(t_0) \exp(ikx_j)$$

a. Find μ_k in terms of a, b, α, dt, dx and k. (Hint: $e^{i\theta} = \cos\theta + i\sin\theta$)

b. For $\alpha=0.5$ and $\alpha=1$, verify that $|\mu_k| \leq 1$ for all k, so that both methods are unconditionally stable.

c. When $a=0$, verify that for Crank-Nicolson $|\mu_k|=1$ for all k, and for the backward difference method min $|\mu_k| = 1 - O((dt/dx)^2)$ and $|\mu_N| = 1$.

d. When $a=Ddx$ (D>0), verify that for both methods, $|\mu_N| = 1 - O(dt/dx)$.

e. You have just shown that adding an artificial viscosity term $Ddx \, u_{xx}$ to the hyperbolic problem $u_t = bu_x$ causes the high frequency noise (k large) to be damped more rapidly. Now calculate the truncation error if the hyperbolic PDE (a=0) is approximated by the finite difference method above with artificial diffusion a=Ddx. Notice that with artificial diffusion the truncation error is no longer $O(dx^2)$ but is still $O(dx)$, so that the finite difference methods still converge to the solution of the hyperbolic problem.

f. Verify that the choice D= $0.5|b|$ results in an "upwind" one-sided difference approximation to the bu_x term. Is it crucial to add this exact amount of artificial diffusion?

5.2 Explain why the eigenvalues of $J\mathbf{w} = \lambda S_C \mathbf{w}$ corresponding to any elastic wave application of the form (1.3.2) will all be real and negative, which guarantees PDE/PROTRAN's stability applied to these problems.

5.3 Consider the homogeneous damped wave equation (cf.1.3.1):

$$u_{tt} + au_t = \nabla^2 u \qquad \text{in R} \qquad a>0 \text{ constant}$$

$$u = 0 \qquad \text{on } \partial R_1$$

$$\partial u/\partial n = 0 \qquad \text{on } \partial R_2$$

$$u, u_t \text{ given at } t=t_0$$

a. Justify the term "damped" by showing that the energy

$$\iint (u_t{}^2 + |\nabla u|^2) \, dxdy$$

is decreasing with time if $a>0$, but is constant if $a=0$.

b. With the aid of the auxiliary variable $v=u_t$, this damped wave equation can be put into the form (5.2.2):

$$u_t = v$$
$$v_t = \nabla^2 u - av$$

Recalculate the stability eigenvalues from 4.2.3 (recall that λ_k must be replaced by ω_k, the square root of λ_k). (Hint: $S_a = aS_I$ since a is constant.) Verify that both the Crank-Nicolson and backward difference methods are still stable, and that the eigenvalues are decreased (in absolute value) by the damping term.

5.4 a. For the example problem (5.1.4) verify that at the endpoints, where u and v may be taken to be 40 and 20, neither of the eigenvectors of F.UX are multiples of u or v alone.

   b. Show that this implies that either u or v can be specified without implicitly specifying either component of

$$P \begin{bmatrix} u \\ v \end{bmatrix}$$

   (Hint: The eigenvectors of F.UX are the columns of P^{-1}.)

*5.5 Use PDE/PROTRAN to solve the transport problem:

$$u_t = -u_x \qquad 0 \leq x \leq 10$$

with boundary condition only at the left endpoint:

$$u(0,t) = 1$$

and initial condition:

$$u(x,0) = \max(1-x,0)$$

The exact solution at time=t has the form:

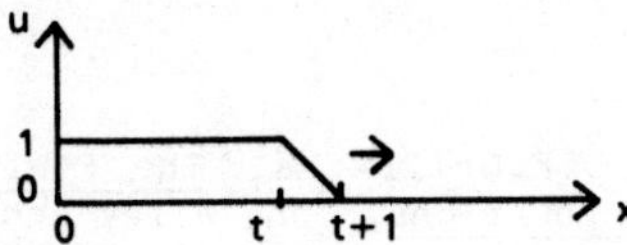

that is, it looks like a train engine (with cowcatcher!) moving from left to right with unit velocity.

Use both the Crank-Nicolson and backward difference methods, and plot the solution (using vertical arrows) at t=3 and t=7. Suggested parameters are DT=0.2, XGRID=(0,0.5,1.0,1.5....10.0), YGRID=(0,1), GRIDPOINTS=(50,0), NOUPDATE. Which of the two time discretization methods better preserves the shape of the solution and why should you have expected this?

You may also want to try this problem with a small amount of artificial viscosity added, even though it is not necessary here, to observe the smoothing effects of diffusion.

*5.6 Use PDE/PROTRAN to solve the wave problem:

$$u_{tt} = u_{xx} \qquad\qquad 0 \le x \le 10$$

with boundary conditions:

$$u(0,t) = 0$$

$$u(10,t) = 0$$

and initial conditions:

$$u(x,0) = \max(1-|x-5|,0)$$
$$u_t(x,0) = 0$$

The exact solution is $u(x,t) = 0.5u(x+t,0) + 0.5u(x-t,0)$. Thus the initial "tent" separates into two smaller tents, which proceed in opposite directions with unit velocity. At time=t, after the halves separate but before they hit the boundary, the solution has the form:

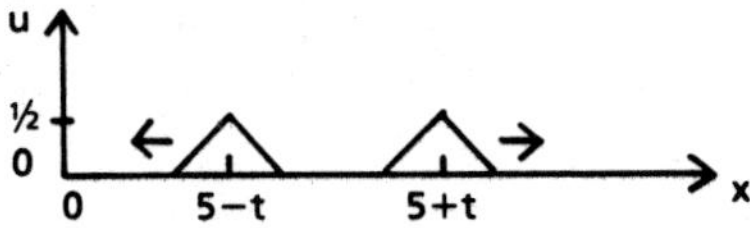

Use both time-discretization methods and plot the solution at t=1 and
t=3.   Suggested parameters are the same as for Problem 5.5.   Which
method better preserves the shape of the tents?

References

1.   A.R. Mitchell and D.F.Griffiths, <u>The Finite Difference Method in
Partial Differential Equations</u>, John Wiley and Sons, New York, (1980).

2.   R.F. Sincovec and N.K. Madsen, "Software for Nonlinear Partial
Differential Equations," <u>ACM Transactions on Mathematical Software,</u> 1,
p245 (1975).

3.   G. Sewell, "IMSL Software for Differential Equations in One Space
Variable," IMSL Tech. Report 8202 (1982).

CHAPTER 6. EIGENVALUE PROBLEMS

6.1 **The Rayleigh-Ritz Approximation**

The form of the eigenvalue PDE system solved by PDE/PROTRAN (Section 1.5) is:

$$0 = \mathbf{A}_x(x,y,\mathbf{u},\mathbf{u}_x,\mathbf{u}_y) + \mathbf{B}_y(x,y,\mathbf{u},\mathbf{u}_x,\mathbf{u}_y) + \mathbf{F}(x,y,\mathbf{u},\mathbf{u}_x,\mathbf{u}_y) + \lambda P(x,y)\mathbf{u} \text{ in R}$$

(6.1.1)
$$\mathbf{u} = \mathbf{0} \qquad \text{on } \partial R_1$$

$$A n_x + B n_y = \mathbf{GB}(x,y,\mathbf{u}) \qquad \text{on } \partial R_2$$

where $\mathbf{A},\mathbf{B},\mathbf{F}$ and $\mathbf{GB}$ are linear, homogeneous functions and P is a (usually diagonal) matrix.

For most values of λ, (6.1.1) will have only the trivial solution $\mathbf{u=0}$. Those values of λ for which a nonzero solution exists are called eigenvalues, and the corresponding solution is called an eigenfunction (see Section 1.4). If any eigenfunction is multiplied by a constant, the result is still a solution to (6.1.1) so the eigenfunctions are not unique.

The eigenfunctions are approximated by functions of the form (cf. 2.1.2 and 4.1.2):

(6.1.2)
$$\mathbf{u_G}(x,y) = \sum_{j=1}^{N} \mathbf{a_j}\phi_j(x,y)$$

where the ϕ_j are the same piecewise polynomial functions as in Section 2.2. Since $\phi_j=0$ on ∂R_1, $\mathbf{u_G}$ satisfies the first (essential) boundary condition automatically.

The "weak formulation" (cf. 2.1.3) of (6.1.1) is obtained, as before, by multiplying the PDE and second boundary condition by an arbitrary smooth function ϕ which satisfies $\phi=0$ on ∂R_1, and integrating over R and ∂R_2. After integration by parts:

(6.1.3) $\iint\limits_{R} (\mathbf{F}\phi-\mathbf{A}\phi_x-\mathbf{B}\phi_y)\ dxdy + \int\limits_{\partial R_2} \mathbf{GB}\phi\ ds = -\lambda \iint\limits_{R} P\mathbf{u}\phi\ dxdy$

And again an approximate eigenfunction of the form (6.1.2) is sought which satisfies the weak formulation only for $\phi = \phi_1,\ldots\phi_N$:

(6.1.4) $\iint\limits_{R}\{-\mathbf{A_G}(\phi_k)_x-\mathbf{B_G}(\phi_k)_y+\mathbf{F_G}\phi_k\}dxdy + \int\limits_{\partial R_2} \mathbf{GB_G}\phi_k\ ds = -\mu \iint\limits_{R} P\mathbf{u_G}\phi_k\ dxdy\ ,$

$$k=1\ldots N$$

113

When (6.1.2) is substituted for u_G in this equation there results an algebraic eigenvalue problem. The left hand side is the same as $f_k(a_1...a_N)$ in (2.1.4) and thus has the same Jacobian (2.1.9). Since A,B,F and GB are linear and homogeneous, this means that f is linear and homogeneous, so the above system reduces to:

$$(6.1.5) \qquad\qquad Ja + \mu S_P a = 0$$

where $a = (a_1...a_N)$, J has elements given by (2.1.9) and S_P has elements given by (4.2.2) with C replaced by P. Recall that the elements J_{kj} and $(S_P)_{kj}$ are really m by m matrices, where m is the number of PDEs.

Thus the PDE eigenvalue system (6.1.1) is approximated by the matrix eigenvalue problem (6.1.5). The solution of this matrix eigenvalue problem is discussed in Section 6.2.

In order to study the error in the eigenvalue approximation, the following functional is defined:

$$H(a,b) = \iint_R (a,a_x,a_y)^T H_1(b,b_x,b_y) \, dxdy + \int_{\partial R_2} a^T H_2 b \, ds$$

where $-H_1$ is the 3m by 3m matrix (2.1.10) and $-H_2$ is just the m by m matrix GB.U. $H(e,e)$ is the same (in the linear case) as defined in Section 2.1, and because of the linearity, H_1 and H_2 are functions of x and y only.

In order to estimate the eigenvalue error, it will be necessary to assume that H_1 and H_2 are symmetric, and thus $H(a,b) = H(b,a)$, and such that $H(e,e)$ is positive for any nonzero e satisfying the essential boundary condition, and that P is positive definite. For linear elastic resonance problems (Section 1.4) this will always be the case since P will be a diagonal density matrix, and $H(e,e)$ is always positive as shown in Section 2.1.

Now since F,A,B and GB are linear and homogeneous:

$$H_1(b,b_x,b_y) = (-F_b,A_b,B_b)$$

$$H_2 b = -GB_b$$

where all functions are evaluated at b.

Thus $H(a,b)$ can alternatively be expressed:

$$H(a,b) = \iint_R (-a^T F_b + a_x^T A_b + a_y^T B_b) \, dxdy - \int_{\partial R_2} a^T GB_b \, ds$$

It will be useful later to note that (6.1.3) is equivalent to requiring that:

(6.1.6) $H(\phi,u) = \lambda \iint\limits_{R} \phi^{T}Pu \; dxdy$

for any vector ϕ satisfying the essential boundary condition. Also notice that (6.1.4) is equivalent to requiring that:

(6.1.7) $H(\phi,u_{G}) = \mu \iint\limits_{R} \phi^{T}Pu_{G} \; dxdy$

for any ϕ in the approximating (piecewise polynomial) set (6.1.2).

Now the Rayleigh-Ritz quotient, which will be useful in estimating the error, is defined as:

$$R(u) = H(u,u) \; / \; \iint\limits_{R} u^{T}Pu \; dxdy$$

If u is an eigenfunction of (6.1.1) it satisfies the essential boundary condition itself, and so (6.1.6) implies that $R(u) = \lambda$.

Similarly, if u_{G} is an eigenfunction of the discrete problem, it is itself a member of the approximating space, and so (6.1.7) implies that $R(u_{G}) = \mu$. Since R is clearly always positive, for nonzero u, all eigenvalues of both (6.1.1) and (6.1.5) are positive.

It will now be shown that the stationary, or critical values, of R correspond to the eigenvalues of (6.1.1). To find the stationary points, let w and ϕ be any two functions satisfying the essential boundary condition. Then for any scalar ε:

$$R(w+\varepsilon\phi) = H(w+\varepsilon\phi,w+\varepsilon\phi) \; / \; \iint\limits_{R} (w+\varepsilon\phi)^{T}P(w+\varepsilon\phi) \; dxdy$$

$$= \{H(w,w)+2\varepsilon H(\phi,w)+\varepsilon^{2}H(\phi,\phi)\} \; / \; \iint\limits_{R} (w+\varepsilon\phi)^{T}P(w+\varepsilon\phi) \; dxdy$$

The stationary points are those functions w for which the derivative of the above expression, evaluated at $\varepsilon=0$, is zero for arbitrary ϕ. This gives, after some simplification:

(6.1.8) $H(\phi,w) = R(w) \iint\limits_{R} \phi^{T}Pw \; dxdy$

Now if $w=u$ is an eigenfunction of (6.1.1), then since $R(u)=\lambda$, (6.1.6) implies that (6.1.8) holds for any ϕ satisfying the essential boundary condition, so that u is a stationary point of R. Conversely, if w is a stationary point of R, (6.1.8) implies that (6.1.6) holds, with eigenvalue $R(w)$.

Similarly, if $\mathbf{w}=\mathbf{u_G}$ is an eigenfunction of the discrete problem (6.1.5), then since $R(\mathbf{u_G}) = \mu$, (6.1.7) implies that (6.1.8) holds for any ϕ in the approximating set. Thus $\mathbf{u_G}$ is a stationary point of R restricted to the approximating set (6.1.2). Conversely, if $\mathbf{w}$ is a stationary point of R restricted to the approximating set, (6.1.8) implies that (6.1.7) holds for any ϕ in this set, so that $\mathbf{w}$ is an eigenfunction of the discrete problem, with eigenvalue $R(\mathbf{w})$.

In particular, since the absolute minimum of R is also a stationary point, the function $\mathbf{u}$ which minimizes R over the space of all functions satisfying the first boundary condition is an eigenfunction of (6.1.1) corresponding to $\lambda_{min}=R(\mathbf{u})$. Similarly, the function $\mathbf{u_G}$ which minimizes R over the approximating space is an eigenfunction of the discrete problem corresponding to $\mu_{min}=R(\mathbf{u_G})$. Thus if $\mathbf{u_I}$ is any other member of the approximating space (e.g. the interpolant to $\mathbf{u}$):

$$(6.1.9) \qquad R(\mathbf{u}) \leq R(\mathbf{u_G}) \leq R(\mathbf{u_I})$$

Now if $\mathbf{e}=\mathbf{u_I}-\mathbf{u}$:

$$R(\mathbf{u_I}) = R(\mathbf{u}+\mathbf{e}) = \{H(\mathbf{u},\mathbf{u}) +2\, H(\mathbf{e},\mathbf{u}) + H(\mathbf{e},\mathbf{e})\} \,/\, \iint_R (\mathbf{u}+\mathbf{e})^T P(\mathbf{u}+\mathbf{e})\ dxdy$$

$$= H(\mathbf{u}+2\mathbf{e},\mathbf{u}) \,/\, \iint_R \{(\mathbf{u}+2\mathbf{e})^T P\mathbf{u} + \mathbf{e}^T P\mathbf{e}\}\ dxdy \ \ + \ \ H(\mathbf{e},\mathbf{e}) \,/\, \iint_R \mathbf{u_I}^T P\mathbf{u_I}\ dxdy$$

Since $\mathbf{u}+2\mathbf{e} = \mathbf{0}$ on ∂R_1, (6.1.6) shows that:

$$H(\mathbf{u}+2\mathbf{e},\mathbf{u}) = \lambda_{min} \iint_R (\mathbf{u}+2\mathbf{e})^T P\mathbf{u}\ dxdy$$

so that:

$$R(\mathbf{u_I}) \leq \lambda_{min} + H(\mathbf{e},\mathbf{e}) \,/\, \iint_R \mathbf{u_I}^T P\mathbf{u_I}\ dxdy$$

Now if the eigenfunction $\mathbf{u}$ is normalized so that $\iint \mathbf{u}^T P\mathbf{u}\ dxdy = 1$ and if $\mathbf{u_I}$ is chosen close enough to $\mathbf{u}$ so that $0.5 \leq \iint \mathbf{u_I}^T P\mathbf{u_I}\ dxdy$ and if an "H-norm" is defined as in Section 2.1 by:

$$||\mathbf{e}||_H^2 = H(\mathbf{e},\mathbf{e})$$

then from (6.1.9):

$$(6.1.10) \qquad \lambda_{min} \leq \mu_{min} \leq \lambda_{min} + 2\, ||\mathbf{u_I}-\mathbf{u}||_H^2$$

Thus (6.1.10) gives a bound on the error in the finite element approximation μ_{min} to the PDE eigenvalue λ_{min}, in terms of the best piecewise polynomial approximation to $\mathbf{u}$. In {1, Section 7.2} similar bounds are given on the error in approximating the higher eigenvalues, which show that the error becomes progressively larger as higher eigenvalues are sought. This is logical, since the PDE problem (6.1.1) has an infinite number of eigenvalues while the matrix eigenvalue problem (6.1.5) has only Nm. Fortunately, it is usually only the smallest, or smallest few, which are of interest in applications.

116

Now if h is the longest edge of any triangle in the triangulation, and n is the degree of the piecewise polynomial approximating functions (n=2,3 or 4 for PDE/PROTRAN), then by Problem 2.13 there exists an approximating function $\mathbf{u_I}$ such that (if $\mathbf{u}$ is smooth):

$$||\mathbf{u_I}-\mathbf{u}||_H^2 = O(h^{2n})$$

and

$$(6.1.11) \qquad \lambda_{min} \leq \mu_{min} \leq \lambda_{min} + O(h^{2n})$$

This means that PDE/PROTRAN, with quartic elements, can achieve $O(h^8)$ accuracy in the eigenvalues! The error in the eigenfunctions is the same as the error in the steady state problem, $O(h^{n+1})$, as shown in {1}. If exact integration is done, (6.1.10) shows that the approximate eigenvalue is always above the true one. However, the fact that numerical integration is done may occasionally alter this order.

It was assumed that H is always positive, but this is not really necessary for (6.1.10) to hold. If the eigenvalues are not all positive, but bounded from below, this bound still holds for the smallest (most negative) eigenvalue (Problem 6.1). Using techniques similar to those employed in Section 4.2, it can be shown that the eigenvalues are bounded from below if the 2m by 2m matrix:

$$\begin{bmatrix} A.UX & A.UY \\ B.UX & B.UY \end{bmatrix}$$

and the matrix P are both positive definite.

To experimentally check the theoretical error bound (6.1.11), the smallest eigenvalue of:

$$\nabla^2 u + \lambda u = 0 \qquad \text{in the unit circle}$$

$$u = 0 \qquad \text{on the boundary}$$

is calculated. The exact value is $\lambda_{min}=5.7831859630$. The error $\mu_{min}-\lambda_{min}$ for all three PDE/PROTRAN elements, with NT (number of triangles) increasing from 20 to 320, is shown in Table 6.1.2. The input used to generate one of these results is shown in Table 6.1.1.

Table 6.1.1

```
C    EIGENVALUE PROBLEM (IN UNIT CIRCLE)
C
$       PDE2D
        NTRIANGLES = 40
        DEGREE = 4
        A = UX
        B = UY
C    LAMBDA*U REPLACED BY U(N)--SEE SECTION 6.2
        F = UN
C    INITIAL GUESS FOR INVERSE POWER METHOD
        UO = 1.-X*X-Y*Y
        CURVES = (-1, DCOS(6.283185307D0*S), DSIN(6.283185307D0*S) )
        NORMALIZE
        NOUPDATE
        SYMMETRIC
        MAXITERATIONS = 10
        VERTICES = (1,0) (-0.5,0.8660254038D0) (-0.5,-0.8660254038D0)
      * (0,0)
        TRIANGLES = (1,2,4,-1)  (2,3,4,-1)  (3,1,4,-1)
        PRECISION = DOUBLE
$       END
```

Table 6.1.2

NT	error using quadratics	error using cubics	error using quartics
20	.7704E-1	.6604E-3	.1455E-3
40	.1389E-1	.2209E-4	.1082E-4
80	.4233E-2	.1322E-4	.4747E-6
160	.8785E-3	.1077E-5	.3877E-6
320	.2835E-3	.2097E-6	.1916E-7

The average slope of log(error) vs log(NT) was -2.02, -2.76 and -3.06 for the three elements. Since, for the uniform triangulations used, NT α h^{-2}, this translates into an error of order 4.04, 5.52 and 6.12 in h, which compares with the order 4,6 and 8 predicted by (6.1.11). Presumably it would require even higher values of NT before the $O(h^8)$ accuracy of the quartic element is observed.

Notice that in all cases, $\mu_{min} > \lambda_{min}$ as predicted by (6.1.10), in spite of the numerical integration.

6.2 **The Inverse Power Method**

Since in applications only the smallest, or smallest few, of the infinite
number of eigenvalues of (6.1.1) are of interest, and since in any case only
the smaller eigenvalues of (6.1.5) are good approximations to eigenvalues of
(6.1.1), methods which simultaneously find all eigenvalues of a matrix
eigenvalue problem are not appropriate here.

Instead, the inverse power method, which finds one eigenvalue and eigenfunction pair at a time of the generalized matrix eigenvalue problem (6.1.5), is
used by PDE/PROTRAN.

For the inverse power method, the user is asked to replace the term $\lambda P\mathbf{u}$ in
(6.1.1) by $P\mathbf{u}^n$ where $\mathbf{u}^n$ is the value of $\mathbf{u}$ on the nth iteration, and then
(6.1.1) is solved exactly as in the steady state case, for the unknown $\mathbf{u}^{n+1}$.

Thus the discrete problem solved each iteration is (see 6.1.5):

$$(6.2.1) \qquad J\mathbf{a}^{n+1} + S_P\mathbf{a}^n = \mathbf{0}$$

$$\mathbf{a}^{n+1} = -J^{-1}S_P\mathbf{a}^n$$

So

$$\mathbf{a}^n = (-J^{-1}S_P)^n\mathbf{a}^0$$

where $\mathbf{a}^n$ is the vector of coefficients of $\mathbf{u}^n$ in the basis expansion (6.1.2).
Now in what follows, it will be assumed that the 3m by 3m matrix H_1 and the
m by m matrix H_2 are symmetric, but not necessarily such that H is positive.
P will still be assumed to be positive definite. Thus J and S_P are symmetric
and S_P is also positive definite. J is assumed to be nonsingular, of
course. The general nonsymmetric case will be discussed briefly later.

Now it is well known that the symmetric eigenvalue problem (6.1.5), with S_P
positive definite, has N linearly independent eigenvectors with real eigenvalues:

$$(6.2.2) \qquad J\,\mathbf{v}_i + \mu_i S_P\mathbf{v}_i = \mathbf{0}$$

where the order of J and S_P is called N. Thus $\mathbf{a}^0$, which is found by requiring
$\mathbf{u}^0$ to interpolate to the user supplied (nonzero) initial values, can be
expanded in an eigenvector basis:

$$\mathbf{a}^0 = \sum_{i=1}^{N} \alpha_i\mathbf{v}_i$$

where (see Problem 6.2) $\alpha_i = \mathbf{v}_i^T S_P\mathbf{a}^0$.

And so:

$$\mathbf{a}^n = \sum_{i=1}^{N} \alpha_i(-J^{-1}S_P)^n\mathbf{v}_i$$

Since the $\mathbf{v}_i$ are eigenvectors of $-J^{-1}S_P$ (see 6.2.2), with eigenvalues $1/\mu_i$:

$$\mathbf{a}^n = \sum_{i=1}^{N} \alpha_i (1/\mu_i)^n \mathbf{v_i}$$

Suppose the eigenvalues are numbered so that $|\mu_i| \leq |\mu_{i+1}|$. Then:

$$(6.2.3) \qquad \mathbf{a}^n = (1/\mu_1)^n \{\alpha_1 \mathbf{v_1} + \sum_{i=2}^{N} \alpha_i (\mu_1/\mu_i)^n \mathbf{v_i}\}$$

Suppose further that μ_1 is a simple eigenvalue, and that $|\mu_1| < |\mu_2|$ (This is usually the case in applications). Then:

$$\lim_{n\to\infty} \mu_1{}^n \mathbf{a}^n = \alpha_1 \mathbf{v_1}$$

Thus unless $\alpha_1 = \mathbf{v_1}^T Sp \mathbf{a}^0 = 0$, which is highly unlikely if $\mathbf{a}^0$ is chosen randomly (and even if it is zero, roundoff error will introduce a nonzero coefficient for $\mathbf{v_1}$ which will grow to be dominant eventually), then for large n, $\mathbf{a}^n$ is approximately proportional to $\mathbf{v_1}$, that is, $\mathbf{a}^n$ is an (unnormalized) eigenvector corresponding to the smallest (in absolute value) eigenvalue μ_1. PDE/PROTRAN will, upon request, normalize $\mathbf{a}^n$ to have largest component of 1. The eigenvalue is estimated by PDE/PROTRAN by:

$$\lim_{n\to\infty} \{(\mathbf{a}^n)^T \mathbf{a}^n\}/\{(\mathbf{a}^n)^T \mathbf{a}^{n+1}\} = \mu_1 \{\alpha_1{}^2 \mathbf{v_1}^T \mathbf{v_1}\}/\{\alpha_1{}^2 \mathbf{v_1}^T \mathbf{v_1}\} = \mu_1$$

Now the rate of convergence to the eigenvalue μ_1 depends on the ratio μ_1/μ_2 where μ_2 is the second smallest eigenvalue. To see this:

$$\lim \ (\mu_2/\mu_1)^n \{(\mathbf{a}^n)^T \mathbf{a}^n/(\mathbf{a}^n)^T \mathbf{a}^{n+1} - \mu_1\} =$$

$$\lim \ (\mu_2/\mu_1)^n \{(\mathbf{a}^n)^T (\mathbf{a}^n - \mu_1 \mathbf{a}^{n+1})\} \ / \ \{(\mathbf{a}^n)^T \mathbf{a}^{n+1}\} =$$

$$\lim \ (\mu_2/\mu_1)^n \mu_1 \{(\mu_1{}^n \mathbf{a}^n)^T (\mu_1{}^n \mathbf{a}^n - \mu_1{}^{n+1} \mathbf{a}^{n+1})\} \ / \ \{(\mu_1{}^n \mathbf{a}^n)^T (\mu_1{}^{n+1} \mathbf{a}^{n+1})\} =$$

$$\lim \ \mu_1 \{(\alpha_1 \mathbf{v_1})^T [\sum_{i=2}^{N} \alpha_i (\mu_2/\mu_i)^n (1-\mu_1/\mu_i) \mathbf{v_i}]\} \ / \ \{(\alpha_1 \mathbf{v_1})^T (\alpha_1 \mathbf{v_1})\} =$$

$$\mu_1 \ \{\alpha_1 \mathbf{v_1}^T (\alpha_2 (1-\mu_1/\mu_2) \mathbf{v_2})\} \ / \ \{\alpha_1{}^2 \mathbf{v_1}^T \mathbf{v_1}\} =$$

$$\mu_1 \ (\alpha_2/\alpha_1)(1-\mu_1/\mu_2)(\mathbf{v_1}^T \mathbf{v_2})/(\mathbf{v_1}^T \mathbf{v_1}) = K$$

It was assumed that $|\mu_2| < |\mu_3|$, but if μ_2 is a multiple eigenvalue this only changes the constant K.

Thus for large n,

$$(6.2.4) \qquad \{(\mathbf{a}^n)^T \mathbf{a}^n\} \ / \ \{(\mathbf{a}^n)^T \mathbf{a}^{n+1}\} = \mu_1 + K(\mu_1/\mu_2)^n$$

Actually, a better estimate of the eigenvalue is given by the "discrete Rayleigh-Ritz" quotient:

$$\{(\mathbf{a}^n)^T Sp \mathbf{a}^n\} \ / \ \{(\mathbf{a}^n)^T Sp \mathbf{a}^{n+1}\}$$

as it can be shown (Problem 6.3) that the error in this case is, for large n, proportional to $(\mu_1/\mu_2)^{2n}$, so that this would cut in half the number of iterations to obtain a given accuracy. Unfortunately, S_P is never explicitly formed by PDE/PROTRAN, and as will be seen shortly, all iterations after the first one are cheap, so that cutting the number of iterations in half does not nearly cut the work in half.

The foregoing shows how the inverse power method obtains an estimate of the eigenvalue of (6.1.5) nearest zero, and the corresponding eigenfunction. However, the inverse power method can also find other eigenvalues, one at a time. If a term $KP\mathbf{u}$ is added to the right hand side of (6.1.1), the eigenvalues of both (6.1.1) and the discrete problem (6.1.5) are shifted by the constant $-K$ and the eigenfunctions are unchanged (see Problem 6.1). Thus:

$$\mu_i{}^{new} = \mu_i{}^{old} - K$$

and now the eigenvalue μ^{new}(closest to 0) of the new problem is equal to μ^{old}(closest to K) - K. Thus the inverse power method may be used to find the smallest eigenvalue of the new problem and the eigenvalue of the original problem nearest K is found simply by adding back the K. This process of finding other eigenvalues has been compared to going fishing--you throw out a K and catch the nearest fish, which may or may not be the one you are looking for.

In addition, since the rate of convergence for the inverse power method applied to the new problem depends on:

$$\frac{\mu^{new}(\text{closest to 0})}{\mu^{new}(\text{next closest to 0})} = \frac{\mu^{old}(\text{closest to K}) - K}{\mu^{old}(\text{next closest to K}) - K}$$

the iteration can be accelerated by throwing K out close to the desired "fish". For example if $\mu_1=9$ and $\mu_2=10$ on the original problem, choosing K=8 will decrease the above ratio from 0.9 to 0.5.

The inverse power method is not limited in its usefulness to symmetric problems, but clearly it cannot find complex eigenvalues, as no complex arithmetic is done. The key to convergence is that there exist a simple, real eigenvalue closer to zero than all the others (see Problem 6.6).

Now the linear system (6.2.1) solved each inverse power iteration has exactly the same matrix, J. Thus NOUPDATE (see Section 4.3) should always be specified and the matrix J will not be updated after the first iteration. If one of the band solvers is used, J is decomposed only on the first iteration, so $O(N^2)$ work will be done then but only $O(N^{1.5})$ work on all subsequent iterations. Since, unlike the time dependent case, the Lanczos method does not have the compensating advantage of a decreased matrix condition number, the direct methods are almost always faster, unless N is astronomically large.

Some other implementations of the inverse power method add $K_n P\mathbf{u}$ to the right hand side of (6.1.1), where K_n is the latest estimate of the eigenvalue. This naturally accelerates the convergence, but has the disadvantage that the matrix must be updated every iteration since K_n is changing. In addition, it is possible for such an iteration to converge to the wrong eigenvalue unless care is taken in the initial stages.

6.3 **Exercises** (* = requires use of PDE/PROTRAN)

6.1 Show that if all eigenvalues of (6.1.1) satisfy $-M < \lambda$, then (6.1.10) still holds, for the smallest (most negative) eigenvalue. (Hint: Add $-MP\mathbf{u}$ to the right hand side of (6.1.1) and show that all eigenvalues of the continuous and discrete problems are translated by $+M$, so that now all are positive and the new H is positive. Then apply (6.1.10) to bound the error in the minimal eigenvalue of the new problem. Finally, eliminate M from this bound.)

6.2 a. Show that the eigenvectors of (6.2.2) are orthogonal in the "S_P inner product", that is, that:

$$\mathbf{v_j}^T S_P \mathbf{v_i} = 0 \qquad \text{when } i \neq j$$

b. Now verify that if the eigenvectors are normalized so that $\mathbf{v_i}^T S_P \mathbf{v_i} = 1$

$$\mathbf{a^0} = \sum (\mathbf{v_i}^T S_P \mathbf{a^0})\mathbf{v_i}$$

6.3 Show that for large n,

$$\mu_1 - (\mathbf{a^n})^T S_P \mathbf{a^n} \, / \, (\mathbf{a^n})^T S_P \mathbf{a^{n+1}} = K(\mu_1/\mu_2)^{2n}$$

(Hint: use Problem 6.2a)

*6.4 Solve
$$\nabla^2 u + \lambda u = 0$$
with $\quad u = 0 \quad$ on the boundary of the region below:

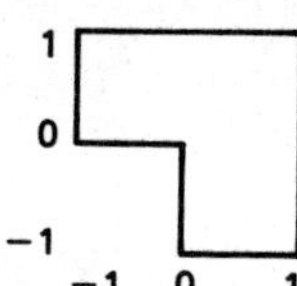

The first two eigenvalues are approximately $\lambda_1 = 9.6$ and $\lambda_2 = 15.2$. Use about 50 cubic triangles, with a triangulation refined locally near the origin, since the eigenfunctions have singularities in their derivatives there {2}.

a. Calculate μ_1 and verify that the inverse power convergence rate is substantially faster than the μ_2/μ_1 predicted. Explain.

b. Verify that adding 9u to the left hand side of the PDE accelerates the inverse power method convergence.

c. Calculate μ_2.

*6.5 The eigenvalues of $\nabla^2 u + \lambda u = 0$ in the square $(0,\pi) \times (0,\pi)$
         with $u = 0$ on the boundary
are:

$$\lambda_{k\ell} = k^2 + \ell^2 \qquad (k=1,2,3\ldots,\ \ell=1,2,3\ldots)$$

Using a fixed triangulation, go fishing for the discrete eigenvalues $\mu_{11}, \mu_{12}=\mu_{21}$ and μ_{22}. Choose the initial values "randomly" since an initial value choice which is almost an eigenfunction for the wrong eigenvalue may lead to convergence (temporarily) to the wrong eigenvalue.

Compare the errors $\mu_{11}-\lambda_{11}$ and $\mu_{22}-\lambda_{22}$. Does the fact that the second eigenvalue is a multiple one affect the inverse power method convergence?

*6.6 Consider the rigid plate eigenvalue problem (see Example 3, Section 1.1) with simply supported boundary:

$$\nabla^2(\nabla^2 u) = \lambda u \qquad \text{in } (0,\pi) \times (0,\pi)$$

$$u = 0 \qquad \text{on the boundary}$$

$$\nabla^2 u = 0$$

a.   Show that the eigenvalues of this problem are the squares of those in Problem 6.5.

b.   Use PDE/PROTRAN to estimate the two smallest eigenvalues. (Hint: introduce an auxiliary variable $v = \nabla^2 u$) Notice that neither J nor S_p in (6.1.5) are symmetric, yet the inverse power iteration converges.

References

1.   R. S. Varga, <u>Functional Analysis and Approximation Theory in Numerical Analysis</u>, Regional Conference Series in Applied Mathematics, SIAM, 1971.

2.   L. Fox, P. Henrici, and C. Moler, "Approximations and Bounds for Eigenvalues of Elliptic Operators," <u>Siam J. Numer. Anal.</u> 4 (1967) pp 89-102.

APPENDIX 1

Lanczos Iteration Properties

Formulas (3.4.2) can be proven by an induction process in which the following are all proven simultaneously:

$$\hat{r}_j^T r_i = 0 \qquad \text{when } i \neq j$$

$$\hat{p}_j^T A p_i = 0 \qquad \text{when } i \neq j$$

$$\hat{p}_j^T A r_i = 0 \qquad \text{when } i < j$$
$$\qquad\quad\; = \hat{p}_i^T A p_i \qquad \text{when } i = j$$

$$\hat{r}_j^T A p_i = 0 \qquad \text{when } i > j$$
$$\qquad\quad\; = \hat{p}_i^T A p_i \qquad \text{when } i = j$$

$$\hat{p}_j^T r_i = 0 \qquad \text{when } i > j$$

For $i=j=0$ the above are trivial, since $p_0=r_0$ and $\hat{p}_0=\hat{r}_0$. Now assume that the above formulas hold for all i and j less than or equal to n. Then it must be proven that they hold for all i and j less than or equal to n+1.

These properties may be illustrated by the boxes:

Box 1 (rows r_i, columns $\hat{r}_j$):

```
        r̂ⱼ
    ┌           ┐
    │ x 0 0 0 │ 0
    │ 0 x 0 0 │ 0
rᵢ  │ 0 0 x 0 │ 0
    │ 0 0 0 xₙ│ 0
    └           ┘
      0 0 0 0  xₙ₊₁
        Box 1
```

Box 2 (rows Ap_i, columns $\hat{p}_j$):

```
        p̂ⱼ
    ┌           ┐
    │ s 0 0 0 │ 0
    │ 0 s 0 0 │ 0
Apᵢ │ 0 0 s 0 │ 0
    │ 0 0 0 sₙ│ 0
    └           ┘
      0 0 0 0  sₙ₊₁
        Box 2
```

Box 3 (rows Ar_i, columns $\hat{p}_j$):

```
        p̂ⱼ
    ┌           ┐
    │ s 0 0 0 │ 0
    │ x s 0 0 │ 0
Arᵢ │ x x s 0 │ 0
    │ x x x sₙ│ 0
    └           ┘
      x x x x  sₙ₊₁
        Box 3
```

Box 4 (rows Ap_i, columns $\hat{r}_j$):

```
        r̂ⱼ
    ┌           ┐
    │ x x x x │ x
    │ 0 s x x │ x
Apᵢ │ 0 0 s x │ x
    │ 0 0 0 sₙ│ x
    └           ┘
      0 0 0 0  sₙ₊₁
        Box 4
```

Box 5 (rows r_i, columns $\hat{p}_j$):

```
        p̂ⱼ
    ┌           ┐
    │ x x x x │ x
    │ 0 x x x │ x
rᵢ  │ 0 0 x x │ x
    │ 0 0 0 xₙ│ x
    └           ┘
      0 0 0 0  xₙ₊₁
        Box 5
```

where the dot products of the indicated vectors are shown, and $s_i = \hat{p}_i^T A p_i$ and x means unspecified. The values inside the solid boxes may be assumed, while those outside are to be proven.

In verifying the values outside the solid boxes above, the formulas in (3.4.1) will be used frequently, as well as the results indicated by the values inside the solid boxes. However, due to the large number of formulas to be verified, verbal explanations will not be given each step of the proofs.

Throughout the proofs below, the letter ℓ is understood to be $\leq n$.

Box 1

$$\tilde{r}_{n+1}{}^T r_\ell = 0$$

Case 1, $\ell=n$:

$$\tilde{r}_{n+1}{}^T r_n = (\tilde{r}_n - \lambda_n A^T \tilde{p}_n)^T r_n$$

$$= \tilde{r}_n{}^T r_n - (\tilde{r}_n{}^T r_n)/(\tilde{p}_n{}^T A p_n)(\tilde{p}_n{}^T A r_n) = 0$$

Case 2, $\ell<n$:

$$\tilde{r}_{n+1}{}^T r_\ell = (\tilde{r}_n - \lambda_n A^T \tilde{p}_n)^T r_\ell$$

$$= 0 - \lambda_n \tilde{p}_n{}^T A r_\ell = 0$$

$$\tilde{r}_\ell{}^T r_{n+1} = 0 \qquad \text{Proof similar to above.}$$

Box 2

$$\tilde{p}_{n+1}{}^T A p_\ell = 0$$

Case 1, $\ell=n$:

$$\tilde{p}_{n+1}{}^T A p_n = (\tilde{r}_{n+1} + \tilde{\alpha}_n \tilde{p}_n)^T A p_n$$

$$= \tilde{r}_{n+1}{}^T A p_n - (\tilde{r}_{n+1}{}^T A p_n)/(\tilde{p}_n{}^T A p_n)(\tilde{p}_n{}^T A p_n) = 0$$

Case 2, $\ell<n$:

$$\tilde{p}_{n+1}{}^T A p_\ell = (\tilde{r}_{n+1} + \tilde{\alpha}_n \tilde{p}_n)^T A p_\ell$$

$$= \tilde{r}_{n+1}{}^T A p_\ell = \tilde{r}_{n+1}{}^T \{(r_\ell - r_{\ell+1})/\lambda_\ell\} = 0$$

$$\tilde{p}_\ell{}^T A p_{n+1} = 0 \qquad \text{Proof similar to above.}$$

Box 3

$$\tilde{p}_{n+1}{}^T A r_\ell = 0$$

Case 1, $\ell=n$:

$$\tilde{p}_{n+1}{}^T A r_n = (\tilde{r}_{n+1} + \tilde{\alpha}_n \tilde{p}_n)^T A r_n$$

$$= \tilde{r}_{n+1}{}^T A r_n - (\tilde{r}_{n+1}{}^T A p_n)/(\tilde{p}_n{}^T A p_n)(\tilde{p}_n{}^T A r_n)$$

$$= \tilde{r}_{n+1}{}^T A (r_n - p_n) = \tilde{r}_{n+1}{}^T A(-\alpha_{n-1} p_{n-1})$$

$$= -\alpha_{n-1} \tilde{r}_{n+1}{}^T \{(r_{n-1} - r_n)/\lambda_{n-1}\} = 0$$

Case 2, $\ell < n$:

$$\tilde{p}_{n+1}{}^T A r_\ell = (\tilde{r}_{n+1} + \tilde{\alpha}_n \tilde{p}_n)^T A r_\ell$$

$$= \tilde{r}_{n+1}{}^T A r_\ell = (\tilde{p}_{n+1} - \tilde{\alpha}_n \tilde{p}_n)^T A (p_\ell - \alpha_{\ell-1} p_{\ell-1})$$

$$= 0$$

$$\tilde{p}_{n+1}{}^T A r_{n+1} = \tilde{p}_{n+1}{}^T A p_{n+1}$$

$$\tilde{p}_{n+1}{}^T A r_{n+1} = \tilde{p}_{n+1}{}^T A (p_{n+1} - \alpha_n p_n)$$

$$= \tilde{p}_{n+1}{}^T A p_{n+1}$$

Box 4

$$\tilde{r}_\ell{}^T A p_{n+1} = 0 \qquad\qquad \text{Proof similar to Box 3}$$

$$\tilde{r}_{n+1}{}^T A p_{n+1} = \tilde{p}_{n+1}{}^T A p_{n+1} \qquad\qquad \text{Proof similar to Box 3}$$

Box 5

$$\tilde{p}_\ell{}^T r_{n+1} = 0$$

Case 1, $\ell = n$:

$$\tilde{p}_n{}^T r_{n+1} = \tilde{p}_n{}^T (r_n - \lambda_n A p_n)$$

$$= \tilde{p}_n{}^T r_n - (\tilde{r}_n{}^T r_n)/(\tilde{p}_n{}^T A p_n)(\tilde{p}_n{}^T A p_n)$$

$$= (\tilde{p}_n - \tilde{r}_n)^T r_n = \tilde{\alpha}_{n-1} \tilde{p}_{n-1}{}^T r_n = 0$$

Case 2, $\ell < n$:

$$\tilde{p}_\ell{}^T r_{n+1} = \tilde{p}_\ell{}^T (r_n - \lambda_n A p_n) = 0$$

This completes the verification of the values outside the solid boxes, and the proof by induction is finished.

APPENDIX 2

The PROTRAN Preprocessor

PDE/PROTRAN is a member of the IMSL PROTRAN family of products, which currently includes MATH/, STAT/, LP/ and PDE/PROTRAN. This means that the PDE/PROTRAN preprocessor has much code in common with the other PROTRAN products, and exhibits the user-friendliness characteristic of all PROTRAN preprocessors.

To illustrate just how flexible and user-friendly PDE/PROTRAN is, consider the following example. Suppose that the elastic crack problem of Section 1.5 (see Table 1.5.1) is to be solved using three different values of the Poisson ratio. The solutions will be stored in PROTRAN arrays. Although these arrays will simply be printed here, they could be processed in any manner desired, using FORTRAN statements.

Since PDE/PROTRAN can be mixed with ordinary FORTRAN in a program, the $PDE2D procedure is placed inside a FORTRAN DO loop, which varies the Poisson ratio, VNU. The initial attempt at solving this problem was done rather clumsily to show off PROTRAN's extensive input error checking facilities. The program, with the resulting error messages, is shown in Table A.2.1.

Once these blunders are corrected, the program is rerun with results as shown in Table A.2.2. This time the preprocessor could detect no errors so it created a FORTRAN program, which was compiled, linked with the PDE/PROTRAN runtime routines, and executed. At execution time, PDE/PROTRAN may issue other warning or error messages to signal numerical difficulties or input errors impossible to detect at preprocessor time. One such message is given here, indicating that the problem is not symmetric, as claimed. Inspection of the equations for A and B reveals the problem--a VY has been incorrectly typed as a UY in the expression for B2.

The final, correct, program is shown in Table A.2.3, with output. Notice that since $PDE2D only requested output on a 5 by 5 grid (GRIDPOINTS keyword), the "ranges" of USOL and VSOL are set to 5 and only the first 5 by 5 portions of the 10 by 10 arrays are printed by the $PRINT procedure. If USOL or VSOL had been dimensioned less that 5 by 5, $PDE2D would have been able to detect this and issue an error message.

Table A.2.1


```
1   C       ELASTIC CRACK PROBLEM
2   C
3   $       DECLARATIONS
4           MATRIX USOL(10,10)
5           DOUBLE PRECISION VECTOR VSOL(10)
6   $       FORTRAN
7   C     SOLVE FOR VNU=0.1,0.2,0.3
8             DO 10 I=1,3
9             VNU = I*0.1
```

```
10   $       PDE2D
11           PRECISION = DOUBLE
12           UNKNOWNS = U,V
13           NTRIANGLES = 50
14           BAND
15           FRONTAL
16           DEGREE = 1
17           A = (E1*UX+E2*VY* , E3*(UY+VX) )
18           B = (E3*(UY+VX) , E1*UY+E2*UX )
19   C   GLOBAL STATEMENT COMMUNICATES VALUE OF VNU TO PDE2D
20           GLOBAL
21           ====
22           COMMON /COMM/ VNU
23           ====
24   C   MATERIAL PROPERTIES DEFINED
25           DEFINED
26           ====
27           EM = 10.6E6
28           E1 = EM/(1.-VNU**2)
29           E2 = E1*VNU
30           E3 = EM/(1.+VNU)/2.
31           ====
32   C   LOCALLY REFINE TRIANGULATION NEAR THE CRACK TIP
33           TRIDENSITY = ((X-1.46)**2+Y**2)**(-1.25)
34   C   SAVE DISPLACEMENTS IN ARRAYS
35           SAVEARRAY = USOL,VSOL
36           GRIDPOINTS = (5,5)
37           SYMETRIC
38   C   NORMALIZE NONUNIQUE DISPLACEMENTS
39           UPRINT = (U-SUBU , V,SUBV)
40   C   APPLY BOUNDARY FORCE ABOVE AND BELOW CRACK OPENINGS
41           GB = (1, 0, -1./2.08)  (6, 0, 1./2.08)
42   C   DEFINE INITIAL TRIANGULATION
43           VERTICES = (0,-2.08) (4.25,-2.08) (4.25,0) (4.25,2.08)
44         *  (0,2.08)  (0,0.001) (0,-0.0001) (2,-1) (1.46,0) (2,1)
45           TRIANGLES = (7,1,8,1) (1,2,8,2) (2,3,8,3) (3,9,8,0) (9,7,8,4)
46         *  (9,3,10,0) (3,4,10,3) (4,5,10,5)  (5,6,10,6)  (6,9,10,7)
47   C   PRINT
```

```
*** FATAL   ERROR.  Invalid PROTRAN keyword:  SYMETRIC

*** FATAL   ERROR.  DEGREE must be between 2 and 4, inclusive.  DEGREE = 1.

*** FATAL   ERROR.  More than one of the following keywords is specified:
***         BAND, FRONTAL, CONJUGATEGRADIENT.

*** FATAL   ERROR.  An arithmetic expression is expected where the
***         following entity specified by keyword A has been used:
***         E1*UX+E2*VY*
```

```
*** FATAL    ERROR.  The required data type is DOUBLE PRECISION where the
***          following entity specified by keyword SAVEARRAY has been used:
***          USOL

*** FATAL    ERROR.  An item with data structure full storage MATRIX is
***          required where the following entity specified by keyword
***          SAVEARRAY has been used: VSOL

*** FATAL    ERROR.  Keyword UPRINT must have exactly 2 items.  The number
***          of items following UPRINT is 3.
   48    $      PRINT VNU,USOL,VSOL
   49           FORMAT = E
   50    $      FORTRAN
   51      10      CONTINUE
   52    $      END
```

Table A.2.2

```
 1    C      ELASTIC CRACK PROBLEM
 2    C
 3    $      DECLARATIONS
 4           DOUBLE PRECISION MATRIX USOL(10,10)
 5           DOUBLE PRECISION MATRIX VSOL(10,10)
 6    $      FORTRAN
 7    C  SOLVE FOR VNU=0.1,0.2,0.3
 8             DO 10 I=1,3
 9             VNU = I*0.1
10    $      PDE2D
11           PRECISION = DOUBLE
12           UNKNOWN = U,V
13           NTRIANGLES = 50
14           BAND
15           DEGREE = 2
16           A = (E1*UX+E2*VY , E3*(UY+VX) )
17           B = (E3*(UY+VX) , E1*UY+E2*UX )
18    C  GLOBAL STATEMENT COMMUNICATES VALUE OF VNU TO PDE2D
19           GLOBAL
20           ====
21           COMMON /COMM/ VNU
22           ====
23    C  MATERIAL PROPERTIES DEFINED
24           DEFINE
25           ====
26           EM = 10.6E6
27           E1 = EM/(1.-VNU**2)
28           E2 = E1*VNU
29           E3 = EM/(1.+VNU)/2
30           ====
31    C  LOCALLY REFINE TRIANGULATION NEAR THE CRACK TIP
```

```
32              TRIDENSITY = ((X-1.46)**2+Y**2)**(-1.25)
33    C  SAVE DISPLACEMENTS IN ARRAYS
34              SAVEARRAY = USOL,VSOL
35              GRIDPOINTS = (5,5)
36              SYMMETRIC
37    C  NORMALIZE NONUNIQUE DISPLACEMENTS
38              UPRINT = (U-SUBU , V-SUBV)
39    C  APPLY BOUNDARY FORCE ABOVE AND BELOW CRACK OPENINGS
40              GB = (1, 0, -1./2.08)   (6, 0, 1./2.08)
41    C  DEFINE INITIAL TRIANGULATION
42              VERTICES = (0,-2.08) (4.25,-2.08) (4.25,0) (4.25,2.08)
43            *  (0,2.08)  (0,0.001) (0,-0.001) (2,-1) (1.46,0) (2,1)
44              TRIANGLES = (7,1,8,1) (1,2,8,2) (2,3,8,3) (3,9,8,0) (9,7,8,4)
45            *  (9,3,10,0) (3,4,10,3) (4,5,10,5)  (5,6,10,6)  (6,9,10,7)
46    C  PRINT DISPLACEMENTS
47    $     PRINT VNU,USOL,VSOL
48          FORMAT = E
49    $     FORTRAN
50       10    CONTINUE
51    $     END
*** WARNING ERROR from $PDE2D at line 10.  Element 6,3 of Jacobian matrix
***           of derivatives of -F1,A1,B1... with respect to U1,U1X,U1Y...
***           appears to be defined nonsymmetrically
```

Table A.2.3

```
1    C      ELASTIC CRACK PROBLEM
2    C
3    $      DECLARATIONS
4           DOUBLE PRECISION MATRIX USOL(10,10)
5           DOUBLE PRECISION MATRIX VSOL(10,10)
6    $      FORTRAN
7    C  SOLVE FOR VNU=0.1,0.2,0.3
8              DO 10 I=1,3
9              VNU = I*0.1
10   $      PDE2D
11          PRECISION = DOUBLE
12          UNKNOWNS = U,V
13          NTRIANGLES = 50
14          BAND
15          DEGREE = 2
16          A = (E1*UX+E2*VY , E3*(UY+VX) )
17          B = (E3*(UY+VY) , E1*VY+E2*UX )
18   C  GLOBAL STATEMENT COMMUNICATES VALUE OF VNU TO PDE2D
19          GLOBAL
20          ====
21          COMMON /COMM/ VNU
22          ====
23   C  MATERIAL PROPERTIES DEFINED
24          DEFINE
```

```
25              ====
26              EM = 10.6E6
27              E1 = EM/(1.-VNU**2)
28              E2 = E1*VNU
29              E3 = EM/(1.+VNU)/2
30              ====
31    C  LOCALLY REFINE TRIANGULATION NEAR THE CRACK TIP
32              TRIDENSITY = ((X-1.46)**2+Y**2)**(-1.25)
33    C  SAVE DISPLACEMENTS IN ARRAYS
34              SAVEARRAY = USOL,VSOL
35              GRIDPOINTS = (5,5)
36              SYMMETRIC
37    C  NORMALIZE NONUNIQUE DISPLACEMENTS
38              UPRINT = (U-SUBU , V-SUBV)
39    C  APPLY BOUNDARY FORCE ABOVE AND BELOW CRACK OPENINGS
40              GB = (1, 0, -1./2.08)  (6, 0, 1./2.08)
41    C  DEFINE INITIAL TRIANGULATION
42              VERTICES = (0,-2.08) (4.25,-2.08) (4.25,0) (4.25,2.08)
43            *  (0,2.08)  (0,0.001) (0,-0.001) (2,-1) (1.46,0) (2,1)
44              TRIANGLES = (7,1,8,1) (1,2,8,2) (2,3,8,3) (3,9,8,0) (9,7,8,4)
45            * (9,3,10,0) (3,4,10,3) (4,5,10,5)  (5,6,10,6)  (6,9,10,7)
46    C  PRINT DISPLACEMENTS
47    $     PRINT VNU,USOL,VSOL
48          FORMAT = E
49    $     FORTRAN
50      10    CONTINUE
51    $     END
```

VNU

.100000E+00

USOL

	1	2	3	4	5
1	.458466E-06	-.184005E-08	-.465951E-06	-.228176E-08	.458345E-06
2	.372240E-06	.372762E-08	-.425569E-06	.324199E-08	.372178E-06
3	.203785E-06	-.192999E-07	-.184477E-06	-.197499E-07	.203259E-06
4	.105240E-06	-.169847E-07	-.996692E-07	-.172649E-07	.104435E-06
5	.742175E-07	-.131382E-07	-.645867E-07	-.133143E-07	.735658E-07

VSOL

	1	2	3	4	5
1	-.926355E-06	-.912239E-06	.927255E-06	.913471E-06	.927574E-06
2	-.387795E-06	-.359641E-06	.339094E-06	.360555E-06	.388509E-06
3	-.745244E-07	-.424283E-07	.754886E-10	.422387E-07	.746827E-07
4	.344791E-07	.321388E-07	.401609E-09	-.308770E-07	-.334650E-07
5.	.981665E-07	.747773E-07	.408670E-09	-.734466E-07	-.965345E-07

PDE/PROTRAN Keywords

PDE2D

Solving PDEs in Two Space Variables

$PDE2D solves the following partial differential equation system:

```
   C(X,Y,T,U)*DU/DT = D/DX*  A(X,Y,T,UX,UY,U)
                     +D/DY*  B(X,Y,T,UX,UY,U)
                     +       F(X,Y,T,UX,UY,U)
```

for (X,Y) in the region R

with   U = FB(X,Y,T)   for X,Y on part of the boundary (∂R_1)

and A*Nx+B*Ny = GB(X,Y,T,U)   for X,Y on the other part (∂R_2), where
                                       (Nx,Ny) is the unit outward normal to
                                       the boundary

and   U = UO(X,Y)   for T=TO

----

where C, A, B, F, FB, GB, UO are user-supplied (possibly vector) functions
of the arguments shown, and U is the solution (vector) and UX, UY represent
(vectors) DU/DX, DU/DY. The product C*DU/DT should be interpreted as the
vector whose i-th component is the product of the i-th components of the
vectors C and DU/DT.

For steady state (time-independent) problems C = 0, and for eigenvalue
problems C = 0 and F may contain terms of the form $\lambda*P(X,Y)*U$.

The user should now divide the boundary of R into distinct arcs, each of
which is smooth with smooth boundary conditions. Thus at every corner or
point where the boundary conditions have a discontinuity or change type, a
new boundary arc must begin. Each arc is given a distinct identifying
integer, I, where I is arbitrary except that it must be positive if the arc
belongs to ∂R_2 (i.e., there are free boundary conditions with GB given
along the arc) and negative if it belongs to ∂R_1 (i.e., there are fixed boundary
conditions with FB given along the arc). Each curved arc is described by a
parameter, s, varying from 0 to 1 (clockwise or counterclockwise).

Next the user creates an initial triangulation of R which generally consists
of only as many triangles as are necessary to satisfy the following rules
(triangles adjacent to a curved boundary may be considered to have one
curved edge):

(1) Each point where two of the boundary arcs meet is included as a
     vertex in the triangulation.

Prohibited:

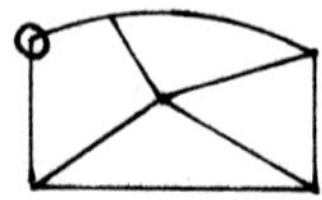

(2) No vertex of any triangle touches another in a point which is not
     a vertex of the second triangle.

Prohibited:

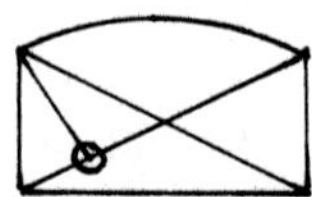

(3) No triangle may have all three vertices on the boundary.

Prohibited:

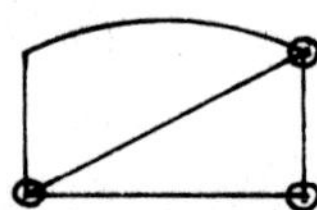

A description of the input now follows.

Procedure reference:

$ PDE2D

Summary of Keyword Phrases:

PRECISION = prec
     If prec = SINGLE or DOUBLE, single or double precision mode is to
     be used.
     The variables in the DEFINE block, and all variables used in
     expressions, are implicitly declared to be REAL or DOUBLE PRECISION
     (depending on prec) if their names begin with A-H or O-Z and
     INTEGER otherwise.
     Default: PRECISION = SINGLE

Partial Differential Equations

NEQUATIONS = neq
     neq is the number of simultaneous partial differential equations
     (1 to 9) in the system to be solved.
     Type: INTEGER
     Structure: Constant
     Default: NEQUATIONS = maximum of the lengths of the lists A,
                 B, F, C

UNKNOWNS = (u1, u2, ..., uneq)

If keyword UNKNOWNS is given, u1, u1X, u1Y, u2, u2X, u2Y, ... may
be used in place of U1, U1X, U1Y, U2, U2X, U2Y... in expressions
where the latter symbols are allowed. u1, u2... must be valid
FORTRAN variables of four or fewer characters, beginning with a
letter in the range A-H or O-Z.

Throughout the documentation for this procedure, if UNKNOWNS is given,
read u1 for U1, u1X for U1X, etc.

In several places the values of the unknowns at the previous time
step or iteration can be used. The components of the previous U
are called U1N, U2N, ..., the components of the previous UX are
called U1XN, U2XN, ..., and the components of the previous UY are
called U1YN, U2YN, ...

Default: UNKNOWNS = U, if neq = 1;
             UNKNOWNS = (U1, U2, ...), if neq > 1.

A = (a1, a2, ..., aneq)
B = (b1, b2, ..., bneq)
F = (f1, f2, ..., fneq)

a1, a2, ..., aneq define the components of A, and
b1, b2, ..., bneq define the components of B, and
f1, f2, ..., fneq define the components of F.

Type: INTEGER, REAL or DOUBLE PRECISION
Structure: Expressions involving constants, X, Y, T, U1, U2,
             ..., U1X, U2X, ..., U1Y, U2Y, ..., U1N, U2N, ...,
             U1XN, U2XN, ..., U1YN, U2YN, ..., KTRI and variables
             defined in DEFINE and GLOBAL.
Default: A = (0, 0, ..., 0)
             B = (0, 0, ..., 0)
             F = (0, 0, ..., 0)

A.UX = (a1.U1X,a1.U2X,...) (a2.U1X,a2.U2X,...) ... (aneq.U1X,aneq.U2X,...)
A.UY = (a1.U1Y,a1.U2Y,...) (a2.U1Y,a2.U2Y,...) ... (aneq.U1Y,aneq.U2Y,...)
A.U = (a1.U1, a1.U2, ...) (a2.U1, a2.U2, ...) ... (aneq.U1, aneq.U2, ...)
B.UX = (b1.U1X,b1.U2X,...) (b2.U1X,b2.U2X,...) ... (bneq.U1X,bneq.U2X,...)
B.UY = (b1.U1Y,b1.U2Y,...) (b2.U1Y,b2.U2Y,...) ... (bneq.U1Y,bneq.U2Y,...)
B.U = (b1.U1, b1.U2, ...) (b2.U1, b2.U2, ...) ... (bneq.U1, bneq.U2, ...)
F.UX = (f1.U1X,f1.U2X,...) (f2.U1X,f2.U2X,...) ... (fneq.U1X,fneq.U2X,...)
F.UY = (f1.U1Y,f1.U2Y,...) (f2.U1Y,f2.U2Y,...) ... (fneq.U1Y,fneq.U2Y,...)
F.U = (f1.U1, f1.U2, ...) (f2.U1, f2.U2, ...) ... (fneq.U1, fneq.U2, ...)

These keywords are required only when the problem is non-linear.

A.UX defines the Jacobian matrix of partial derivatives of (a1, a2,
..., aneq) with respect to (U1X, U2X, ...). Thus, for example, aI.UJX
represents the partial derivative of aI with respect to UJX, and is
a function of the same arguments as aI. It defaults correctly provided
aI is a linear function of UJX. All the elements of A.UX are defaulted

135

if the keyword A.UX is omitted. To default individual elements within
the Jacobian matrix, assign a value of * to the element.
Type: INTEGER, REAL or DOUBLE PRECISION
Structure: Expression involving constants, X, Y, T, U1, U2, ..., U1X,
               U2X, ..., U1Y, U2Y, ..., U1N, U2N, ..., U1XN, U2XN,
               ..., U1YN, U2YN, ..., KTRI and variables defined in
               DEFINE and GLOBAL.
Default: a̲I.UJX defaults correctly provided a̲I is a linear function
               of UJX, that is, provided a̲I.UJX is̲ independent of UJX.
               Otherwise do not default.

DEFINE
  ====
     A block of FORTRAN code to compute parameters used by the PDE
     coefficient keywords A, B, F, A.UX, A.UY, A.U, B.UX, B.UY, B.U,
     F.UX, F.UY, F.U, C and INTEGRAL.
  ====
     This block should be viewed as a segment to do preliminary calculations
     for the above expressions, and only the variables which these keywords
     may reference can used in this FORTRAN block.
GLOBAL
  ====
     A block of FORTRAN containing declaration statements (e.g. COMMON,
     REAL, DOUBLE PRECISION, INTEGER). These statements are inserted into
     the main program and each subroutine generated by PROTRAN. This allows
     the value of a variable defined in the main PROTRAN program to be
     available to any expression, provided it is mentioned in a COMMON
     statement in the GLOBAL block.
  ====

UO = (u̲1, u̲2, ..., u̲neq)
     u̲1, u̲2, ..., u̲neq define the components of UO. For steady state and
     eigenvalue problems, UO gives the initial values for the iterative
     method (Newton's method or inverse power method, respectively)
     used. For linear steady state problems the initial values are
     usually defaulted to zero.
     Type: INTEGER, REAL or DOUBLE PRECISION
     Structure: Expressions involving constants, X, Y, KTRI and variables
                    defined in GLOBAL.
     Default: UO = (0, 0, ..., 0)

SYMMETRIC
     If this keyword is given it is assumed that the following two Jacobian
     matrices are symmetric, at all points:

$$
\begin{bmatrix}
F.U & F.UX & F.UY \\
-A.U & -A.UX & -A.UY \\
-B.U & -B.UX & -B.UY
\end{bmatrix}
$$

     and

          GB.U

Note that each element of the above two matrices is an <u>neq</u> by <u>neq</u>
matrix, so both may be thought of as block matrices. A warning message
is issued at execution time if SYMMETRIC is given for a nonsymmetric
problem. If SYMMETRIC is not given for a symmetric problem, the storage
and execution time are greater than necessary.
Default: The problem is assumed to be nonsymmetric.

<u>Initial Triangulation</u>

If the region R is a rectangle, bounded by exactly four boundary arcs X=XA,
X=XB, Y=YA, Y=YB, then the region and initial triangulation may be defined using
the keywords XGRID, YGRID, ARCS. Otherwise the keywords CURVES, VERTICES,
TRIANGLES must be used.

    XGRID = <u>x1</u>, <u>x2</u>, ..., <u>xm</u>
    YGRID = <u>y1</u>, <u>y2</u>, ..., <u>yn</u>

        <u>x1</u>, <u>x2</u>, ..., <u>xm</u> are the x coordinates of the vertical grid lines.
        <u>y1</u>, <u>y2</u>, ..., <u>yn</u> are the y coordinates of the horizontal grid lines.
        The initial triangulation generated by these grid lines will have four
        equal area triangles in each grid square, so the total number of
        triangles will be 4*(m-1) *(n-1). It is assumed that <u>x1</u>=XA, <u>xm</u>=XB,
        <u>y1</u>=YA, <u>yn</u>=YB.
        Type: INTEGER, REAL or DOUBLE PRECISION
        Structure: Expressions involving constants and variables defined
            in GLOBAL.
    ARCS = (<u>ixa</u>, <u>ixb</u>) (<u>iya</u>, <u>iyb</u>)
        <u>ixa</u>, <u>ixb</u> are the arc numbers of the rectangle edges X=XA,X=XB
        respectively, and <u>iya</u>, <u>iyb</u> are the arc numbers of the edges Y=YA, Y=YB.
        Type: INTEGER
        Structure: Expressions involving constants and variables defined
            in GLOBAL.

    CURVES = (<u>iarc1</u>, <u>x1</u>, <u>y1</u>) (<u>iarc2</u>, <u>x2</u>, <u>y2</u>) ...
        This keyword gives the parametric equations for all curved boundary
        arcs. X=<u>xI</u>(S),Y=<u>yI</u>(S) are the parametric equations which trace out
        boundary arc number <u>iarcI</u> as S varies from 0 to 1.
        Type: <u>iarcJ</u> are INTEGER,
              <u>xJ</u>, <u>yJ</u> are INTEGER, REAL or DOUBLE PRECISION
        Structure: Expressions involving constants, S, and variables defined
            in GLOBAL.
         Default: All boundary arcs not mentioned are assumed to be straight
            lines.

    VERTICES = (<u>x1</u>, <u>y1</u>) (<u>x2</u>, <u>y2</u>) ...
        (<u>xI</u>, <u>yI</u>) are the coordinates of vertex number I of the initial triangu-
        lation. The vertices may be listed in any order, but that order defines
        the vertex numbers referred to in the TRIANGLES keyword. Each vertex
        is listed only once.
        Type: INTEGER, REAL or DOUBLE PRECISION
        Structure: Expressions involving constants and variables defined
            in GLOBAL.

137

TRIANGLES = (ia1, ib1, ic1, iarc1) (ia2, ib2, ic2, iarc2) ...
    iaI, ibI, icI are the numbers (see VERTICES keyword) of the vertices
    A, B, C of triangle I of the initial triangulation. A, B, C must
    be ordered counterclockwise and such that C is not on the boundary.
    iarcI is the arc number of the boundary arc intercepted by the
    base, AB, of triangle I. If AB is in the interior of the region
    put iarcI=0.
    Type: INTEGER
    Structure: Expressions involving constants and variables defined
           in GLOBAL.

    Note: Some expressions may contain a reference to the integer parameter
    KTRI, which at any point (X,Y) holds the number of the triangle in
    the initial triangulation in which (X,Y) lies (triangle numbers
    are determined by the order in the TRIANGLES keyword). This feature
    is useful in defining problems in a composite region.

One Dimensional Problems
    A problem involving only one space variable, with (say) $X1 \leq X \leq X2$,
    can be handled relatively efficiently by PDE/PROTRAN.   Such a problem
    can be considered to be a two dimensional problem (with all functions
    independent of Y) on the rectangle $X1 \leq X \leq X2$, $0 \leq Y \leq 1$, with boundary
    conditions at Y = 0 and Y = 1 determined from the fact that the normal
    derivatives of all unknowns are zero there.

    The grid option should be used to define the initial triangulation;
    automatic triangulation refinement using NTRIANGLES need not be
    used. XGRID should be used to specify the grid points for the one-
    dimensional problem and two grid lines in the Y direction should be
    specified using YGRID = (0,1). The storage and execution time increase
    linearly with the number of grid lines (if BAND or FRONTAL is specified)
    in the X direction, since the Jacobian bandwidth remains approximately
    constant. By contrast, for a two dimensional problem the execution
    time increases as the fourth power of the number of grid lines in
    the X or Y direction.

Triangulation Refinement

NTRIANGLES = ntf
    ntf is the number of triangles desired in the final triangulation.
    ntf must be greater than or equal to the number of triangles in the
    initial triangulation.

    If $h = \text{NTRIANGLES}^{**}(-1/2)$ and DEGREE = ndeg is the element degree
    then the error in U for elliptic problems is proportional to
    $h^{**}(\text{ndeg}+1)$, even when the boundary is curved. If the mesh is
    properly graded (using TRIDENSITY) this error rate holds even for
    singular problems. The error in A and B is proportional to $h^{**}\text{ndeg}$.
    For time-dependent (parabolic) problems there is also a time-discre-
    tization error in addition to this spacial discretization error.
    See the discussion after the keywords CRANKNICOLSON and BACKWARDS
    for details.

Type: INTEGER
Structure: Constant
Default: NTRIANGLES = MAX (50, itri), where itri is the number
             of triangles in the initial triangulation.

TRIDENSITY = d3est
    d3est is a function of X and Y which controls the grading of the
    triangular grid in the final triangulation. The initial triangulation
    is refined with the goal of distributing d3est $(X,Y)**(2/3)*A(j)$
    evenly over the final triangulation, where A(j) is the area of
    triangle j. Normally the user simply makes d3est largest where the
    final triangulation is to be most dense and then plots it (see
    PRINTERPLOT) to see if it is graded satisfactorily.
    Type: INTEGER, REAL or DOUBLE PRECISION
    Structure: Expression involving constants, X, Y, KTRI and variables
                 defined in GLOBAL.
    Default: TRIDENSITY = 1.0 (a uniform triangulation)

SHAPE = shape
    shape is a function of X and Y which controls the approximate shape
    of the triangles in the final triangulation. The initial triangulation
    is refined with the goal of generating triangles with an average height
    (delta y) to width (delta x) ratio of shape(X,Y) near the point (X,Y).
    Type: INTEGER, REAL or DOUBLE PRECISION
    Structure: Expression involving constants, X, Y, KTRI and variables
                 defined in GLOBAL.
    Default: SHAPE = 1.0 (approximately equilateral triangles)

PRINTERPLOT
    If given, printer plots of the initial triangulation (for input
    checking) and of the centers and vertices of the triangles in the
    final triangulation will be generated.
    Default: No printer plots are generated.

Boundary Conditions

FB = (iarc1, fb1, fb2, ...) (iarc2, fb1, fb2, ...) ...
    fb1, fb2, ... define the components of FB on arc number iarcI
    (iarcI must be negative).
    Type: iarcI are INTEGER,
                 fbJ are INTEGER, REAL or DOUBLE PRECISION
    Structure: iarcI are expressions involving constants and variables
                 defined in GLOBAL.
                 fbJ are expressions involving constants, X, Y, T, S (on
                 curves, the arc parameter) and variables defined in GLOBAL.
    Default: FB = (0, 0, ..., 0) on all arcs with negative numbers.

GB = (iarc1, gb1, gb2, ...) (iarc2, gb1, gb2, ...) ...
    gb1, gb2, ... define the components of GB on arc number iarcI (iarcI
    must be positive).
    Type: iarcI are INTEGER,
                 gbJ are INTEGER, REAL or DOUBLE PRECISION

Structure: iarcI are expressions involving constants and variables
defined in GLOBAL.
gbJ are expressions involving constants, X, Y, T, S (on
curves, the arc parameter), U1, U2, ... U1N, U2N... and
variables defined in GLOBAL.

Default: GB = (0, 0, ..., 0) on all arcs with positive numbers.

GB.U = (iarc1, gb1.U1, gb1.U2, ...) (iarc1, gb2.U1, gb2.U2, ...) ...
      (iarc2, gb1.U1, gb1.U2, ...) (iarc2, gb2.U1, gb2.U2, ...) ...

        .   .   .   .

        .   .   .   .

This keyword is required only when the problem is non-linear.

GB.U defines the Jacobian matrix of partial derivatives of (gb1,
gb2, ...) with respect to (U1, U2, ...), on each arc. Thus, for
example, gbJ.UK represents the partial derivative of gbJ with
respect to UK (on arc number iarcI) and is a function of the same
arguments as gbJ. It defaults correctly provided gbJ is a linear
function of UK. All the elements of the Jacobian matrix are defaulted
on an arc if that arc is not mentioned. To default individual
elements within the Jacobian matrix, assign a value of * to the element.

Type: iarcI are INTEGER,
gbJ.UK are INTEGER, REAL or DOUBLE PRECISION

Structure: iarcI are expressions involving constants and variables
defined in GLOBAL.
gbJ.UK are expressions involving constants, X, Y, T, S
(on curves, the arc parameter), U1, U2, ... U1N, U2N... and
variables defined in GLOBAL.

Default: For all arcs not mentioned, it is assumed that gb1,
gb2, ... are linear functions of U1,U2... and these
derivatives are calculated correctly. Otherwise do not
default.

Mixed Boundary Conditions
Boundary conditions of different types on the same arc cannot be
directly specified. However, a boundary condition of the type u =
FB(s,t) can be replaced by a boundary condition of the type

$$A*n_x + B*n_y = -\beta*(u-FB(s,t))$$

where β is a very large number.

Periodic Boundary Conditions
Periodic boundary conditions can be handled by rewriting the PDE.
The following example illustrates how periodic boundary conditions
may be treated.

Solve $\rho_{xx} + \rho_{yy} + f(x,y) = 0$ on a rectangle with $\rho(x,0) = g_0(x)$,
$\rho(x,1) = g_1(x)$, $\rho(-\pi,y) = \rho(\pi,y)$, $\rho_x(-\pi,y) = \rho_x(\pi,y)$.

Defining the new variables
$$u(x,y) = \rho(x,y) - \rho(-x,y)$$
$$v(x,y) = \rho(x,y) + \rho(-x,y)$$
the differential equations become (on $(0,\pi)$ x $(0,1)$)
$$u_{xx} + u_{yy} + f(x,y) - f(-x,y) = 0$$
$$v_{xx} + v_{yy} + f(x,y) + f(-x,y) = 0$$
and the boundary conditions may be defined on the reduced region as

$$u(x,0) = g_0(x) - g_0(-x) \qquad v(x,0) = g_0(x) + g_0(-x)$$

$$u(x,1) = g_1(x) - g_1(-x) \qquad v(x,1) = g_1(x) + g_1(-x)$$

$$u(0,y) = 0 \qquad v_x(0,y) = 0$$

$$u(\pi,y) = 0 \qquad v_x(\pi,y) = 0$$

The boundary conditions along x=0 and x=π may be written

$$\left(\frac{\partial u}{\partial n}, \frac{\partial v}{\partial n}\right) = GB = (\beta u, 0)$$

where β is very large. It is suggested that UPRINT be used to
cause $\rho(x,y) = (u+v)/2$ and $\rho(-x,y) = (v-u)/2$ to be output in place
of u and v.

<u>Output</u>

GRIDPOINTS = (<u>nx</u>, <u>ny</u>)
    The solution is output on an <u>nx</u> by <u>ny</u> uniform grid.
    Type: INTEGER
    Structure: Constants
    Default: GRIDPOINTS = (9, 9)

GRIDLIMITS = (<u>xa</u>, <u>xb</u>) (<u>ya</u>, <u>yb</u>)
    The solution is output on a uniform grid covering the rectangle
    <u>xa</u> $\leq$ x $\leq$ <u>xb</u>, <u>ya</u> $\leq$ y $\leq$ <u>yb</u>.

    If output is desired at an arbitrary sequence of points, rather than
    on a uniform rectangular grid, set

        GRIDLIMITS = (<u>x</u>(M),<u>x</u>(M)) (<u>y</u>(M),<u>y</u>(M))

    where <u>x</u>(M),<u>y</u>(M) are expressions which are functions of M and (<u>x</u>(M),
    <u>y</u>(M)), M=1...nx*<u>ny</u> are the desired points.
    Type: INTEGER, REAL or DOUBLE PRECISION
    Structure: Expressions involving constants, M, and variables
            defined in GLOBAL.
    Default: GRIDLIMITS = (xmin, xmax) (ymin, ymax)
            where (xmin, xmax) (ymin, ymax) define the smallest
            rectangle containing the domain.

OUTFREQUENCY = <u>nout</u>
    The solution is output every <u>nout</u> time steps or iterations.
    Type: INTEGER
    Structure: Expression involving constants and variables defined in

                         GLOBAL.
          Default: OUTFREQUENCY = 1

PRINTSOLUTION
     If given, a table of solution values is printed.
     Default: The solution is not printed.

SAVEFILE = filename
     If given, the solution is written onto file 'filename'. If this
     solution is to be later processed by PLOTSURFACE, PLOTSTRESSES or
     PLOTVECTOR, note that the default filename for the input file for
     these procedures is 'PLOT'. Filenames beginning with 'IMSL' are
     reserved and should not be used. This keyword acts independently
     of PRINTSOLUTION.
     Default: The solution is not written to a file.

SAVEARRAY = (array1, array2, ...)
     arrayI is an nx by ny matrix to hold the I-th component of the solution
     vector at the output points. arrayI(i,j) is assigned the value of
     UI (possibly as modified by the UPRINT keyword) at the output point
     (X(i),Y(j)). If the problem is time-dependent or iterated, the solution
     at the last time value or iteration is stored in arrayI. If (X(i),Y(j))
     lies outside the region R, arrayI(i,j) is assigned an extrapolated
     value.
     Type: As specified by PRECISION
     Structure: MATRIX
     Default: The solution is not stored in an array.

SAVEFUNCTION = (fname1, fname2, ...)
     fnameI is the name to be given a FORTRAN function subprogram which
     can be called after execution of $PDE2D to evaluate the Ith component
     of the solution vector. fnameI(X,Y) returns the value of UI (possibly
     as modified by the UPRINT keyword) at the point (X,Y). If the problem
     is time-dependent or iterated the solution at the last time value
     or iteration is returned.

     This keyword should only be given if GRIDLIMITS is defaulted. If (X,Y)
     is not a point of the output grid, an interpolation is done, and a
     serious loss of accuracy may result if the grid is coarse or the
     solution ill-behaved. If (X,Y) lies outside the region R, fnameI(X,Y)
     is assigned an extrapolated value.
     Type: As specified by PRECISION
     Structure: SCALAR
     Default: No solution functions are created.

INTEGRAL = int
     The integral (over R) of expression int is calculated and output along
     with the solution.
     Type: INTEGER, REAL or DOUBLE PRECISION
     Structure: Expression involving constants, X, Y, T, U1X, U2X, ...,
                  U1Y, U2Y, ..., U1, U2, ..., KTRI and variables defined
                  in DEFINE and GLOBAL.

UPRINT = (u1print, u2print, ...)
APRINT = (a1print, a2print, ...)
BPRINT = (b1print, b2print, ...)
     Normally U1, U2, ..., A1, A2..., B1, B2, ... are printed at the output
     points. The above keywords can be used to redefine these output
     variables. For example, u1print defines the variables to be output
     in place of UI.

     Certain PDE problems, such as Laplace's equation with Neumann boundary
     conditions only, do not have a unique solution. To aid in the
     normalization of this solution the variables U1AV, U2AV, ...,
     containing the average values of U1, U2, ... over the region R, are
     defined.

     In the important case of a plane strain or plane stress elasticity
     problem with only boundary forces given and all forces independent
     of U and V, the displacements are unique only to within a trans-
     lation and a rotation. The variables SUBU and SUBV are calculated
     internally in such a way that the displacement field U-SUBU, V-SUBV
     has an average displacement of zero and a net rotational moment of
     zero.
     Type: INTEGER, REAL or DOUBLE PRECISION
     Structure: Expression involving constants, X, Y, T, KTRI, U1, U2,
                  ..., A1, A2, ..., B1, B2, ..., A, B, U1AV, U2AV, ...,
                  SUBU, SUBV, and variables defined in GLOBAL.

     Default: UPRINT = (U1,U2...)
                  APRINT = (A1,A2...)
                  BPRINT = (B1,B2...)

NORMALIZE
     If present, the problem is assumed to be an eigenvalue problem and
     the eigenfunction is normalized each iteration to have maximum value
     of one.
     Default: The eigenfunction is not normalized.

COMPARE = filename
     If this keyword is specified, it is assumed that an earlier run has
     been made which differs from the current run only in the element degree
     used, and that the solution from that problem is stored on file
     'filename'. The maximum change over the output points is calculated
     and output automatically to provide an estimate of the space
     discretization error.
     Default: No compare is done.

Solution Method

DEGREE = ndeg
     ndeg is the degree of the element to be used. Quadratic (DEGREE=2),
     cubic (DEGREE=3) and quartic (DEGREE=4) isoparametric elements are
     available.
     Type: INTEGER

```
Structure:  Constant
Default:    DEGREE = 2
```

BAND
FRONTAL
CONJUGATEGRADIENT

These keywords specify the linear equation solution method to be
employed:

```
BAND -      A band solver is used.  This is generally the fastest
            method for small and moderate size problems, but requires
            more storage than the other two methods.
FRONTAL -   An out-of-core band solver (frontal method) is used.
            This is typically 50% slower than the in-core band
            solver, but requires much less storage.
CONJUGATEGRADIENT -  A preconditioned bi-conjugate gradient
            (Lanczos) method is used.  This is generally the best
            choice for large elliptic and parabolic problems, but
            it may occasionally fail to converge.  DOUBLE PRECISION
            is recommended if this method is used.
Default:    BAND
```

NDIMENSION = ndim

ndim is the number of words of storage to be allocated for the large
Jacobian (stiffness) matrix. NDIMENSION should normally be defaulted.
However, for the BAND and FRONTAL solvers it is impossible to predict
the exact amount of storage needed a priori, so an error message may
occasionally be issued at execution time telling the user that the
default storage is not sufficient and suggesting a value for NDIMENSION
to override the default.

```
Type:       INTEGER
Structure:  Constant
Default:    If DEGREE = 2
            and BAND method used:  12*ntf**1.5*neq**2*nsym
            and FRONTAL method:    20*ntf      *neq**2*nsym
            and CONJUGATE method:  36*(ntf+8) *neq**2
                                    +(neq+8)*(3*ntf+1)

            If DEGREE = 3
            and BAND method used:  60*ntf**1.5*neq**2*nsym
            and FRONTAL method:    100*ntf      *neq**2*nsym
            and CONJUGATE method:  100*(ntf+8) *neq**2
                                    +(neq+8)*(6*ntf+1)

            If DEGREE = 4
            and BAND method used:  192*ntf**1.5*neq**2*nsym
            and FRONTAL method:    320*ntf      *neq**2*nsym
            and CONJUGATE method:  225*(ntf+8) *neq**2
                                    +(neq+8)*(10*ntf+1)

            where ntf = NTRIANGLES
                  neq = NEQUATIONS
                  nsym = 1 if SYMMETRIC given
                       = 2 if SYMMETRIC not given
```

MAXITERATIONS = <u>niter</u>

    <u>niter</u> is an upper limit on the number of iterations of Newton's method to be done to solve a steady state problem, or number of iterations of the inverse power method to be done to solve an eigenvalue problem. For linear steady state problems MAXITERATIONS is normally defaulted to one.

    The Newton's method iteration stops if convergence occurs before <u>niter</u> iterations.

    Type: INTEGER
    Structure: Expression involving constants and variables defined in GLOBAL.
    Default: MAXITERATIONS = 1

STEPLIMIT

    If present, Newton's method is modified so that the change in the solution each iteration is less than 30% of the maximum value of the solution on the previous iteration. This usually increases the probability that the Newton iteration will converge.
    Default: The step size is not limited.

NOUPDATE

    This keyword may be given if a constant time step is used, the problem is linear and all functions, except inhomogeneous terms, are independent of time. In other words, NOUPDATE may be specified if C and all keywords of the form A.B are functions of X, Y and S only and if DTDENSITY is defaulted. All eigenvalue problems and many parabolic problems fall into this category. If NOUPDATE is not specified and the message 'NOTE-JACOBIAN UPDATED UNNECESSARILY' appears each step after the first, then NOUPDATE may be specified the next time the problem is run.

    If NOUPDATE is specified a large savings in execution time is realized. It causes the LU decomposition of the Jacobian matrix (or the Jacobian itself, if the keyword CONJUGATEGRADIENT is given) computed at the first time step or iteration to be saved and used on all subsequent time steps or iterations.

    Default: The Jacobian is updated every time step or iteration (unless AUTOUPDATE is specified).

<u>Time Dependent Problems</u>

    C = (<u>c1</u>, <u>c2</u>, ..., <u>cneq</u>)
        <u>c1</u>, <u>c2</u>, ..., <u>cneq</u> define the components of C.
        Type: INTEGER, REAL or DOUBLE PRECISION
        Structure: Expressions involving constants, X, Y, T, U1, U2, ..., KTRI and variables defined in DEFINE and GLOBAL.
        Default: C = (0, 0, ..., 0)

    TO = <u>t0</u>
        <u>t0</u> is the initial time value.

145

Type: INTEGER, REAL or DOUBLE PRECISION
Structure: Expression involving constants and variables defined in
             GLOBAL.
Default: TO = 0.0

TF = tf
     tf is the final time value.
     Type: INTEGER, REAL or DOUBLE PRECISION
     Structure: Expression involving constants and variables defined in
                  GLOBAL.
     Default: TF = 1.0

DT = dt
     dt is the average time step size.
     Type: INTEGER, REAL or DOUBLE PRECISION
     Structure: Expression involving constants and variables defined in
                  GLOBAL.
     Default: DT = 1.0

DTDENSITY = dtinv
     dtinv is a function of T which controls the time step variation (the
     average time step size is still dt). At time T, the step size is
     proportional to 1/dtinv(T).
     Type: INTEGER, REAL or DOUBLE PRECISION
     Structure: Expression involving constants, T and variables defined
                  in GLOBAL.
     Default: DTDENSITY = 1.0 (a constant time step)

CRANKNICOLSON
BACKWARDS
     These keywords specify the time discretization method to be used.
     CRANKNICOLSON (Crank-Nicolson) is a second order stable scheme and
     BACKWARDS (implicit backwards difference) is a first order stable
     scheme.

     If BACKWARDS is used then the error due to the time-discretization
     is proportional to dt (the time step). If CRANKNICOLSON is used
     then the error is proportional to dt^2. If EXTRAPOLATION is also
     used the errors are then proportional to dt^2 and dt^4. The error in
     the solution U is proportional to the sum of the time-discretization
     and spacial-discretization errors. See NTRIANGLES for a discussion
     of the spacial-discretization errors.
     Default: BACKWARDS

AUTOUPDATE
     If given, the frequency of updating of the Jacobian is determined
     adaptively. AUTOUPDATE should not be specified if the CONJUGATEGRADIENT
     solution method is used.
     Default: The Jacobian is updated every step (unless NOUPDATE
                  is specified).

EXTRAPOLATE = <u>filename</u>
>   If this keyword is specified it is assumed that an earlier run has
been made which differs from the current run only in that the average
time step size (<u>dt</u>) was exactly half of what it is in the current
run, and that the solution from that problem is stored on file
'<u>filename</u>'. A Richardson extrapolation is done using the results
from the previous run and the current run, and the extrapolated
results are printed in place of the current solution. The order of
the time discretization error in the extrapolated values is twice
the order in the unextrapolated values.

>   An estimate of the time discretization error is also calculated and
output. Recall, however, that the total error also includes a space
discretization error, which depends on the mesh spacing. It can be
estimated using COMPARE.
Default: No extrapolation is done.

PLOTSTRESSES

Generating a Stress Plot

$PLOTSTRESSES generates a plot of the principal stresses for a solution calculated and saved by $PDE2D. It is assumed that A1,B2,A2 = B1 are respectively the x and y tensile stresses and the shear stress.

Procedure reference:

$ PLOTSTRESSES

Summary of Keyword Phrases:

    PLOTFILE = <u>filename</u>
        '<u>filename</u>' is the name of the file where the $PDE2D tabular output
        has been stored.
        Default: PLOTFILE = PLOT

    SET = <u>nset</u>
        <u>nset</u> is the number of the solution set on file '<u>filename</u>' which is
        to be plotted. Thus, for example, if the solution had been output
        with $PDE2D parameters DT = 0.1, OUTFREQUENCY = 10 and the solution
        at T = 3.0 is desired, <u>nset</u> should be 3.
        Type: INTEGER
        Structure: Expression
        Default: SET = 1

    NODISTORT
        If given, the x and y axes are scaled equally, so that the region is
        undistorted.
        Default: The x and y axes are scaled so that the figure approximately
                    fills the plot space.

    TITLE = '<u>title</u>'
        '<u>title</u>' is a character string specifying the plot title.
        A maximum of 40 characters are allowed.
        Default: No title is written.

148

PLOTSURFACE

Generating a Surface Plot

$PLOTSURFACE generates a surface plot of a solution calculated and saved by
$PDE2D.

Procedure reference:

$ PLOTSURFACE

Summary of Keyword Phrases:

    PLOTFILE = filename
        'filename' is the name of the file where the $PDE2D tabular output
        has been stored.
        Default: PLOTFILE = PLOT

    SET = nset
        nset is the number of the solution set on file 'filename' which is
        to be plotted. Thus, for example, if the solution had been output
        with $PDE2D parameters DT = 0.1, OUTFREQUENCY = 10 and the solution
        at T = 3.0 is desired, nset should be 3.
        Type: INTEGER
        Structure: Expression
        Default: SET = 1

    SURFACE = variable1
        A surface plot is to be made of variable1, where variable1 is one
        of the symbols:

        U1,A1,B1,U2,A2,B2....-U1,-A1,-B1,-U2,-A2,-B2...U,A,B,-U,-A,-B

        Default: SURFACE = U1

    VIEWLATITUDE = lat
    VIEWLONGITUDE = long
        lat and long are the latitude and longitude of the view angle, in
        degrees.

        Both angles must be between 10 and 80 degrees.
        Type: INTEGER, REAL or DOUBLE PRECISION
        Structure: Expression
        Default: VIEWLATITUDE = 45.0
                    VIEWLONGITUDE = 45.0

    TITLE = 'title'
        'title' is a character string specifying the plot title.
        A maximum of 40 characters are allowed.
        Default: No title is written.

149

Generating a Vector Field Plot

$PLOTVECTOR generates a vector field plot of a solution calculated and saved by $PDE2D.

Procedure reference:

$ PLOTVECTOR

Summary of Keyword Phrases:

    PLOTFILE = filename
         'filename' is the name of the file where the $PDE2D tabular output
         has been stored.
         Default: PLOTFILE = PLOT

    SET = nset
         nset is the number of the solution set on file 'filename' which is
         to be plotted. Thus, for example, if the solution had been output
         with $PDE2D parameters DT = 0.1, OUTFREQUENCY = 10 and the solution
         at T = 3.0 is desired, nset should be 3.
         Type: INTEGER
         Structure: Expression
         Default: SET = 1

    VECTOR = (variable1,variable2)
         The vector (variable1,variable2) is to be plotted with arrows indicating
         magnitude and direction. variable1 and variable2 may be chosen from
         the symbols:

         0,U1,A1,B1,U2,A2,B2...-U1,-A1,-B1,-U2,-A2,-B2...U,A,B,-U,-A,-B

         If variable1=0, variable2 will be plotted with vertical arrows
         representing magnitude.
         Default: VECTOR = (U1, U2)

    NODISTORT
         If given, the x and y axes are scaled equally, so that the region
         is undistorted.
         Default: The x and y axes are scaled so that the figure approximately
                      fills the plot space.

    TITLE = 'title'
         'title' is a character string specifying the plot title.
         A maximum of 40 characters are allowed.
         Default: No title is written.

Generating a Contour Plot

$PLOTCONTOUR generates a contour plot of a solution calculated and saved by $PDE2D.

Procedure reference:

    $ PLOTCONTOUR

Summary of Keyword Phrases:

    PLOTFILE = filename
        'filename' is the name of the file where the $PDE2D tabular output
        has been stored.
        Default: PLOTFILE = PLOT

    SET = nset
        nset is the number of the solution set on file 'filename' which is
        to be plotted. Thus, for example, if the solution had been output
        with $PDE2D parameters DT = 0.1, OUTFREQUENCY = 10 and the solution
        at T = 3.0 is desired, nset should be 3.
        Type: INTEGER
        Structure: Expression
        Default: SET = 1

CONTOUR = variable1
        A contour plot is to be made of variable1, where variable1 is one
        of the symbols:

        U1,A1,B1,U2,A2,B2...-U1,-A1,-B1,-U2,-A2,-B2,...U,A,B,-U,-A,-B

        Default: CONTOUR = U1

NODISTORT
        If given, the x and y axes are scaled equally, so that the region
        is undistorted.
        Default: The x and y axes are scaled so that the figure approximately
                    fills the plot space.

TITLE = 'title'
        'title' is a character string specifying the plot title.
        A maximum of 40 characters are allowed.
        Default: No title is written.

APPENDIX 4

Postscript

As I begin my 12th year of work on TWODEPEP (now PDE/PROTRAN), I am intrigued by
the analogy between the 11 year evolution of this computer code and the
multi-billion year history of the genetic code of life, which contains a blueprint
for a species encoded into billion of bits of information. Like the code of
life, TWODEPEP began with primitive features, being capable of solving only a
single linear elliptic equation in polygonal regions, with simple boundary
conditions. It passed through many useful stages as it adapted to non-linear and
time dependent problems, systems of PDEs, eigenvalue problems, and as it evolved
cubic and quartic elements and isoparametric elements for curved boundaries. It
grew a preprocessor and a graphical output package, and out-of-core frontal and
conjugate gradient methods were added to solve the linear systems.

Each of these changes represented major evolutionary steps--new orders, classes
or phyla, if you will. The conjugate gradient method, in turn, also passed
through several less major variations as the basic method was modified to precon-
dition the matrix, to handle nonsymmetric systems, and as stopping criteria were
altered, etc. Some of these variations might be considered new families, some
new genera, and some only special changes.

I see one flaw in the analogy, however. While I am told that the DNA code was
designed by a natural process capable of recognizing improvements but incapable
of planning beyond the next random mutation, I find it difficult to believe that
TWODEPEP could have been designed by a programmer incapable of thinking ahead
more than a few characters at a time.

But perhaps, it might be suggested, a programmer capable of making only random
changes, but quite skilled at recognizing improvements could, given 4.5 billion
years to work on it, evolve such a program. A few simple calculations would
convince him that this programmer would have to rely on very tiny improvements.
For example, if he could produce a billion random "mutations" per second (or, for
a better analogy, suppose a billion programmers produce one "mutation" per second
each), he could not, statistically, hope to produce any predetermined 20 character
improvement during this time period. Could such a programmer, with no programming
or mathematical skills other than the ability to recognize and select out very
small improvements through testing, design a sophisticated finite element program?

The Darwinist would presumably say, yes, but to anyone who has had minimal
programming experience such an idea is preposterous. The major changes to
TWODEPEP, such as the addition of a new linear equation solver or new element,
required the addition or modification of hundreds of lines of code before the new
feature was functional. None of the changes made during this period were of any
use whatever until all were in place.

Even the smallest modifications to that new feature, once it was functional,
required adding several lines, no one of which made any sense, or provided any
"selective advantage", when added by itself.

Consider, by way of analogy, the airtight trap of the carnivorous bladderwort plant, which has a double sealed, valve-like door which is opened when a trigger hair is activated, causing the victim to be sucked into the vacuum of the trap (described by R.F.Daubenmire in "Plants and Environment," John Wiley and Sons, N.Y. 1947). It is difficult to see what selective advantage this trap provided until it was almost perfect.

This, then, is the fallacy of Darwin's explanation for the causes of evolution--the idea that major (complex) improvements can be broken down into many minor improvements. French biologist Jean Rostand, in "A Biologist's View" (William Heinemann, Ltd., London, 1956) recognized this:

> "It does not seem strictly impossible that mutations should have introduced into the animal kingdom the differences that exist between one species and the next...hence it is very tempting to lay also at their door the differences between orders, classes and phyla--in short, the whole of evolution. But it is obvious that such an extrapolation involves the gratuitous attribution to the mutations of the past of a magnitude and power of innovation much greater than is shown by those of today."

The famous "problem of novelties" is another formulation of the objection raised here. How can natural selection cause new organs to arise and guide their development through the initial useless stages during which they present no selective advantage, the argument goes. The Darwinist is forced to argue that there are no useless stages. He believes that new organs and new systems of organs arose gradually, through many small improvements. But this is like saying that TWODEPEP could have made the transition from a single PDE to systems of PDEs through many five or six character improvements, each of which made it work slightly better on systems.

It is interesting to note that this belief is not supported even by the fossil evidence. Harvand paleontologist George Gaylord Simpson, for example, in "The History of Life," Volume II of "Evolution after Darwin," (University of Chicago Press, 1960) points out:

> "It is a feature of the known fossil record that most taxa appear abruptly. They are not, as a rule, led up to by a sequence of almost imperceptably changing forerunners such as Darwin believed should be usual in evolution...This phenomenon becomes more intense and more universal as the hierarchy of categories is ascended. Gaps among known species are sporadic and often small. Gaps among known orders, classes and phyla are systematic and almost always large. These peculiarities of the record pose one of the most important theoretical problems in the whole history of life: Is the sudden appearance of higher categories a phenomenon of evolution or of the record only, due to sampling bias and other inadequacies?"

Another way of describing this same structure is expressed in a recent _Life_ magazine article (Francis Hitching, "Was Darwin Wrong on Evolution?", April 1982, which concludes that "natural selection has been tested and found wanting") which focuses on the "curious consistancy" of the fossil gaps:

"These are not negligible gaps. They are periods, in all the major
evolutionary transitions, when immense physiological changes had to take
place."

Unless we are willing to believe that useless, "developing" organs (and insect
traps which could almost catch insects) abounded in the past, we should have
expected the fossil structure outlined above, with large gaps between the higher
categories, where new organs and new systems of organs appeared.

Nevertheless, despite the fact that the structure of the fossil record is the
only argument against Darwin which has received much attention lately, this is
not the real issue. The "problem of novelties" correctly states the real argument,
but too weakly. Consider, for example, the human eye, with an aperture whose size
varies automatically according to the light intensity, controlled by reflex
signals from the brain; with a lens whose curvature varies automatically according
to the distance to the object in view; and with a retina which receives the
picture on color sensitive cells and transmits it, complete with coded intensity
and frequency information, through the optic nerve to the brain. The brain
superimposes the pictures from the two eyes and stores this 3D picture somehow in
memory, and it will be able to search for and recall this image later and use it
to recognize an older but familiar face in a different picture. Like TWODEPEP,
the eye has passed through various useful stages in its development, but it
contains a large number of features which could not reach usefulness in a single
random mutation and which provided no selective advantage until useful (e.g. the
nerves and arteries which service it), and many groups of features which are
useless individually. The Darwinist may bridge the gaps between taxa with a long
chain of tiny improvements in his imagination, but the analogy with software puts
his ideas into perspective. The idea that all the magnificent species in the
living world, or the human brain with its human consciousness, could have arisen
from simple organic molecules guided by a natural process unable to plan beyond
the next tiny mutation, is entirely comparable to the idea that a programmer
incapable of thinking ahead more than a few characters at a time could, given a
lot of time, design any sophisticated computer program.

I suggest that, with Jean Rostand, "we must have the courage to recognize that we
know nothing of the mechanism" of evolution.